Executive Engineering

Jack Danger

Copyright © 2024 Jack Danger

All rights reserved.

ISBN: 979-8-9897616-1-6

Contents

Preface 1

Engineering Output 3
    How Engineering and Product Deliver Together 11
    The Shipped Potential Chart 26
    Maximizing Output 28
    The Fundamental Forces of Engineering 30

Technical Coherence: A Theory of Engineering 35
    Infrastructure Gravity & Domain Engineering 44
    Step 1: Identifying UX Domains 63
    Step 2: Identify Shared Domains 72
    Step 3: Staffing the Breadth and Depth of Engineering 77

Technical Debt Financing 91
    Common Pitfalls of Addressing Technical Debt 96
    Calculating Technical Debt 102

Empowered Teams 119
    Focused Experience Teams 123
    Competency-Based Agile 138
    Deliberate Engineering 150
    Communication Design 159

This book is dedicated to all of the engineers and engineering managers who don't look impressive on a promotion spreadsheet but spend every day working hard on the right thing for their teammates and their company.

You're my superheroes.

Preface

I've written this book for CTOs who need the most efficient performance from their team. I'll use "CTO" to mean whoever is responsible for all engineers and technology at a company. CTOs have many responsibilities but this book focuses very narrowly on exactly one: **delivering real value faster**. Because if we can't produce and deliver value to users then we lose the right to do the rest of the job.

What follows is the distillation of the solutions I've learned to one of software's hardest problems: How to increase shipping[1] speed continuously over time despite a constrained financial runway. That is, how to make each engineer on each team deliver more value to users year over year.

These solutions aren't tips or tricks — they're a reckoning of the fundamental forces that exist within engineering. Just as organic chemistry undergirds medicine, as aeronautics is foundational to aviation, there are principles underneath software engineering that we cannot escape.

We'll investigate those principles and see how as CTO we have significant control over the shipping speed through the contexts we set. And how, as we lower the drag our team faces, their creative output increases without bound.

I'm approaching a quarter of a century of building SaaS startups. I've had every leadership role from CEO (not recommended), CTO, head of Product, VP, Director, EM, and spent plenty of time as an IC engineer (highly recommended) from entry-level to Staff to Principal. I've worked at companies of every size and funding round, from being a founder and first engineer up to leading teams at Square through its IPO.

I've got a sense for where a company trips itself up, having been partly responsible for many stumbles myself. And, this might be as surprising to you as it was to me, I find that the technical problems in large engineering orgs are often identical in shape to problems at tiny startups – just blown up across more code and data.

---

[1] I use "shipping" to mean the entire cycle of understanding customer needs, designing a technological solution, building it, securing it, putting it into the customer's hands, guaranteeing its correctness, and fixing it when it fails.

Those problems are often hidden by high revenue or VC funding. If money is no constraint then you can match ballooning problems with ballooning headcount – adding extra people to obfuscate the rough edges of the technical and managerial strategies. We don't always have that luxury, so this book is about leading our teams to success quickly, affordably, and happily.

The mental models and values in this book are the same ones that I use every day in my own hands-on, technical executive leadership. Implementing the lessons in this book requires having enough authority to set policies so I've aimed the book at the person individually responsible for the function of Engineering, but the principles hold true regardless of our current roles. I hope that whatever your current (or next) job is, these can help you and your team get where you're going.

How to read this book

You'll notice some words are capitalized, like Engineering, and Product. When I'm referring to a function – an entire subset of an org chart – at a company I will capitalize it. This is to distinguish between the general idea of doing, say, marketing with the Marketing function at a company, which has specific deliverables.

For parity with CTO I'll use CPO as a shorthand for Chief Product Officer, even if the person leading Product is called something different.

Part One

Engineering Output

Velocity

Ensuring Survival

Your company might run out of money.

This is a foundational truth of business and a reason for your CEO's restless nights. Your job, as CTO, is to help remove your CEO's worry and to ensure your engineers take the least time to deliver the maximum value.

Some lucky startups do everything wrong and still have revenue. Trying to copy their internal processes is useful only if we've also got their luck. For most of us, at most companies, even doing everything right may end with us closing our doors. Therefore every executive needs to ensure their function of the company costs less than the value it delivers. If we can't do this, we're not yet operating at the executive level; we're just a really senior manager.

Doing this is hard for executives of any function but it's super hard for whoever leads Engineering. The work that Engineering does only results in revenue after it passes through many layers of abstraction and ultimately through Sales, Marketing and Customer Support. So even identifying the value that Engineering provides is a herculean task, to say nothing of trying to increase it.

And yet, that's the job. To ensure Engineering is delivering for the company and to prove that to some degree. Enough for the CEO and your peer executives to not worry if you've got this handled.

You might be tempted to borrow the habits of large, rich, established companies here. If you have a monopoly on hardware or ad publishing or defense contracts then it's best to use the practices entrenched at Apple, Google, and Microsoft, respectively. If not, you'll need to be much more careful about understanding what you're doing, what you're not doing, how it connects to the business plan, and how it's all going at any given time.

CEO Empathy

We start with the CEO. Your boss. A modern CEO hears countless messages from venture capitalists and from other CEOs about moving faster. There's advice about firing not just your low-performers but firing the regular folks too.[2] There's advice so vague that it sounds right but CEOs could do literally anything and believe they're implementing it.[3] And, just to make this situation impossible, Marc Andreesson himself says that great founders "[d]on't take any advice."[4] Some of this is useful but much of the advice out there has no effect other than to make the person saying it look smart and make everyone hearing it feel anxious.

I hear all advice from VCs in the context of VC impatience. They don't get their money back until the company has a liquidation event like an IPO or a sale. Until then a VC can either help or shout encouraging words about going faster. Few of them can help so we hear a lot of the latter.

The more junior your CEO is the more they may feel this external pressure to rush forward, but no CEO is immune to it. And since they don't know how, specifically, to make Engineering go faster this pressure lands on your shoulders in a vague, anxious way, at best. At worst, your CEO starts to get involved in the details of your org and directs day-to-day activities but without the decades of eng leadership experience necessary to do it right.

Which is why they need you for this. The CEO probably can't (and definitely shouldn't have to) figure out whether Engineering is on track. Your job as CTO differs from the role of an engineering manager or director in this key way: It's on you to free the CEO from worrying about this whole function of the company.

---

[2] Gokul Rajaram, who might not be wrong: https://www.linkedin.com/posts/gokulrajaram1_ceos-of-early-stage-startups-the-most-dangerous-activity-7090823824643932160-AZZN

[3] "Hire the right people, don't run out of money, and have a north star." — Rob Hayes https://review.firstround.com/Heres-the-Advice-I-Give-All-of-Our-First-Time-Founders

[4] https://steno.ai/podcast/lex-fridman-podcast-10-transcript/386-marc-andreessen-future-of-the-internet-technology-and

Proving that you're delivering

To support your CEO you must first make sure Engineering is working fast enough on the right things. Then you have to prove that to the CEO. This is very hard. You can easily fail by either focusing on mere image management so the CEO *thinks* you're delivering while Engineering languishes, or by focusing just on the actual work you think is important while failing to prove your success to the CEO. In both cases you've introduced misalignment.

You can only survive this misalignment and a lack of real velocity metrics for as long as company revenue exceeds costs and there's no real market competition. But the moment the CEO worries about Engineering output you're in trouble. Now you need to make changes —some changes, of some kind — until your work passes the CEO's vibe-check, something that's impossible to do with any professionalism at all. You could launch countless products and show the CEO detailed roadmaps and get into long discussions about process latencies and, well, it might work. Or you might get fired. Or something in between. After all, you haven't provided the CEO a meaningful measurement. Just vibes.

You'll need a real metric the moment your CEO uses the term 'velocity' or, specifically, 'engineering velocity'. It'll have to be a good metric, too, because you're about to have to double or triple it. So, while we all want to talk about how to actually make Engineering faster, we must first face the realities of optics and measuring Engineering velocity.

The smartest CTOs I know don't tolerate conversations about "velocity." Not "engineering velocity" or "product velocity" — they redirect the conversation to make sure there's a clear business strategy, then a clear product strategy to implement the business strategy, and then a technology strategy[5] in turn. This redirection from "velocity" to actual goals gives an executive team real problems to fix that lead to real outcomes.

Conversations about "velocity" fail regularly and sometimes spectacularly. If they lead to new initiatives there's a high likelihood that, instead of speeding up engineering work, it just slows things down, adding a layer of bullshit that in no way helps the company deliver more value.

---

[5] https://lethain.com/eng-strategies/

And yet *we* need to know how fast we're going. Even if we redirect the larger conversation away from 'velocity' we're the executive who sets the engineering pace. We own the whole portfolio of cost, value, and risk for our function. In the most practical terms, our job is to make the entire Engineering function something that doesn't cause the CEO worry. And a part of that is being able to prove that the right work is happening at the fastest reasonable pace.

We'll need to grapple with 'velocity' one way or another, so let's do it right. I'm about to spend several chapters discussing how to do it well but first lets look at how it usually goes wrong.

There are some easily-measured engineering processes, like time spent on a project or the story points in tickets. These represent costs to the company because the ticket points represent estimated time spent, which approximates to salary paid by the company. These measurements are so easy to chart along a timeline that most initiatives to increase velocity just use these. But these are cost metrics, not value metrics. Consultants invented story points to get the client to agree in advance that a feature might take a while so they pay up when it finally does take that long. Measuring cost helps one get a check to recoup those costs.

We don't have clients, though, only cross-functional teammates. And we want a measure of value — the CEO already knows how much engineering costs, it appears every month in the bank statements.

There are two other common patterns for measuring velocity we should dispense with; the first is measuring velocity through OKRs.[6] Ever since John Doerr brought a watered-down version of Andy Grove's management style[7] to Google in the 1990s, Silicon Valley has repeatedly grasped at OKRs as a way to organize all of our teams' work. OKRs translate goals between different levels of abstraction. That's better than nothing but a team sets OKRs based on what they believe they can do, regardless of how much is actually needed. This metric could look great right up until the company runs out of money and shuts its doors.

---

[6] "Objectives and Key Results" are a good way to define a goal and define a way to measure progress. Sub-orgs can define sub-goals in the same way, spanning an entire company.

[7] J. Doerr, Measure What Matters: OKRs: The Simple Idea that Drives 10x Growth. Baltimore, MD, USA: Portfolio, 2017.

There's one more pattern that an old colleague of mine calls "Managing to Weird Shit." It's where we pick some internal engineering process and pretend it's representative of the value delivered, like deploy times, code review latency, chat messages sent or received, etc. Using some part of the engineering process as if it's correlated with valuable outcomes immediately runs afoul of Goodhart's Law[8] and actually slows down the work.

Much work has gone into defining metrics that are valuable within Engineering, to engineers. The DORA metrics,[9] for example, help a team track the resilience of their system. The SPACE[10] framework is a great pulse-check for leadership in precisely the same way a doctor might want to know a patient's pulse. But virtually none of it is useful outside Engineering. To a CEO, internal velocity metrics are as useful as measuring the quality of a doctor by how high they could drive a patient's pulse.

And yet we need a metric.

We need to prove to ourselves, our CEO, and our peers that Engineering is delivering for the company and doing so quickly. We need a metric that incentivizes the actual work Engineering should be doing and maximizes the output that we want from this function.

This metric exists, and it's one that captures the real value delivered from Engineering. The reason you don't see it more often is because it requires an alignment between the executives leading Product and Engineering (and maybe Design) on some precise definition of their roles. It also requires buy-in from the executive team to agree about what the company has historically accomplished.

To surface this metric we'll need to look much closer at the actual role of Engineering within the company.

---

[8] "Any observed statistical regularity will tend to collapse once pressure is placed upon it for control purposes."
[9] https://cloud.google.com/blog/products/devops-sre
[10] https://queue.acm.org/detail.cfm?id=3454124

How Engineering and Product Deliver Together

Where Engineering sits in the value chain

The company must maximize value while minimizing cost. It's a pretty simple formula: Total revenue in dollars divided by all expenses, including salaries. Aim to maximize this ratio and you'll be aligned with your CEO and the board.

Engineering drives this ratio higher with technology-leveraged experiences. Automation that does for (nearly) free what a human would have to do laboriously. You know, software.

Many functions of the company work together to make this software and make it valuable to users. Each function has a distinct role.

- **Product researches, defines and coordinates opportunities** for the company

- **Design constructs a low-friction user experience** that melds the new opportunities with all historical business decisions and user experiences

- **Engineering implements the opportunities that are the highest value and lowest cost** in the context of the existing technological landscape

- **Sales and Marketing shepherd users to the experiences** and experiences to the users, respectively

- **Operations bridges the gap** for all user experiences that were either not automated or not foreseen.

Product and Engineering Relationships

I've heard the role of CPO described as "The CEO of the product" more times than I can count. The number of people trying to break into this role is enormous and, anecdotally, many of them seem to resonate with that description. Being the "CEO of the product" makes it sound like this person is the decision-maker for features, driving products forward. Something like this:

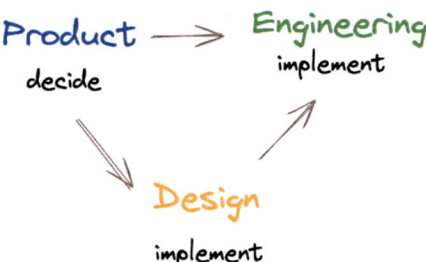

While being a mini-CEO sounds empowering it's often just the opposite: You feel responsible for the product outcome so you get all the anxiety without the power of the real CEO. I've seen CPOs spend long hours using soft influence on adjacent orgs and hard, direct influence on designers and engineers.

Not only is this kind of product leadership exhausting to the person in the role, it degrades the teamwork between Product, Engineering and Design. Instead of three distinct functions working as peers, there's one function attempting to oversee the work of all three.

This delivers sub-par work no matter which function leads. When the founders of a company are engineers they may view technical work as the boss function, giving Engineering more power over defining the work than other functions. At most companies it's the other way around, the founders may fall into the trap of delegating "What should we build?" entirely to the head of Product, empowering them to direct the work of Engineering and Design.

These failure modes are so predictable we can describe the mess that comes from each one.

When Engineering gets ahead of Product

A junior CTO in charge

- The CTO may unconsciously act as the architect, setting themselves (or a delegated 'architect', typically their friend) as the bottleneck for engineering work

- Results in weird investments that in no way contribute to the competitive advantage of the company

- A pursuit of what's interesting, novel, or satisfying to individual engineers over what's useful for the business

A junior CTO in charge with many engineering leaders

- Can lead to a proliferation of tools and architectures, any one of which would have been good if it were the *only* one selected

- No ability to calibrate engineers to career levels because the teams have different cultures and are working in different, incomparable ways

- The product roadmap has a competing, alternate engineering roadmap

- Engineering teams may find ways to work on something more interesting — but not more important — than product features

A senior CTO in charge

- Sets a balance of product shipping and debt paydown with a focus on engineer happiness

- A competent CPO is likely to leave because of the undervaluing of Product, leading to the CTO working with junior product folks such that everyone involved is pretty far from understanding the market

- These companies are often lovely places for engineers to work, right up until a much more focused competitor steals the whole market from them

When Product gets ahead of Engineering

A junior CPO in charge

- The work is dominated by urgent tasks, flowing directly from the CEO's anxiety

- Reactive to small changes in customer feedback and process breakdowns

- Roadmap is regenerated frequently but rarely usable, putting Engineering and Product planning at odds

A junior CPO in charge with lots of other Product teammates

- Overwhelming noise coming out of the Product org

- Meetings that decide product direction are too large and too slow to include a quorum of engineers so decisions are made without sufficient Engineering leadership involvement

- Engineering may ignore Product to work on deprioritized work important to engineers

A senior CPO in charge

- Product planning and execution become the only highly-visible work at the company, leading to under-focus on security, maintainability, performance, or (ironically) development velocity

- Engineering may be unable to protect the hidden layers of work underneath Product Engineering, leading to costly misses in platform and infrastructure work

- Inability to retain senior engineers who want better technical culture

- Debt that is invisible to Product can accumulate until Engineering spends all its time keeping the lights on

The Agency Model: The Original Startup Sin

The above failure modes all have an imbalance between executive teammates and they really boil down to just two causes. When Engineering gets ahead of Product it's because the company leadership doesn't care enough to solve real user problems. They'd rather just build something cool. Congratulations, you're definitely going to get bought by a competitor for cheap.

Or, and this is far more common, Product gets ahead of Engineering because the company leadership thinks that that's how features get built. That someone dreams up what they want and then gets Engineering to make it.

Product telling Engineering what to build is not a relationship between two powerful professionals trying to maximize company value, it's the relationship between an eccentric millionaire and a development consultancy. It's a model that uses in-house engineers as if they're an external development agency.

Development consultants at agencies develop features with the singular goal of "make the client happy enough to write a check" but in a product engineering company we have other meaningful goals. And there's nobody writing a check. If an eccentric millionaire backs themselves into a corner with shortsighted decisions it's their own dollar on the line but in a company every stakeholder pays the price.

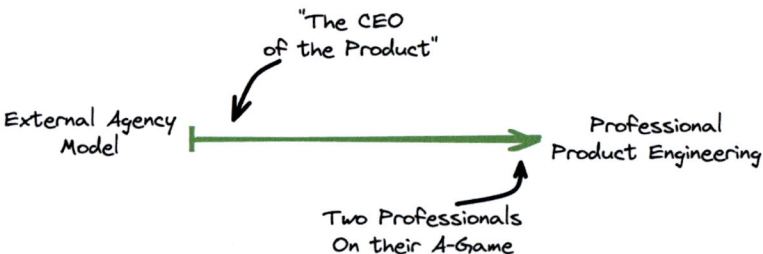

There's a place for the agency model within a company, like when Security or Legal act as consultants to help other teams with specialized problems. For the actual development of the product we need something stronger, something that can fully utilize every skilled professional.

Product Engineering: Peers, not Bosses

To find that stronger model, let's look a little closer at one specific handoff between Engineering and Product: Sequencing the work.

Engineering must own technical sequencing decisions.

Engineering may not have the ability to make the first pass at reasonable sequencing but they need to have the final say. Engineering may feel a strong temptation to delegate this to Product and Product may feel a strong desire to do it, but this has to be Engineering's responsibility because sequencing requires comprehensive knowledge of the entire set of options including technical and staffing nuances that are only available to Engineering.

Deciding what gets built in which order isn't just choosing what features get launched. It's also choosing which technical migrations to start, pause, and finish. Some projects are cheaper after a new tool is available or after a key engineer is freed up from other work. Optimal sequencing requires deep knowledge of the technical system, business opportunities, and the growth trajectories of engineers.

So Engineering must own sequencing decisions and must collaborate tightly with Product to understand what is being sequenced. This powerful relationship between Product and Engineering allows Product to identify opportunities and communicates them in terms of the defined product strategy, with a defined product value, and likely with a *recommended* sequence. Engineering then adjusts the sequence so as to quickly deliver both the maximum quantity of product value as well as a software system that maintains its integrity over time, allowing development to accelerate.

The EPD Model

As each function grows more into fully understanding their own work rather than defining the work of others, a powerful collaboration emerges. And it finally makes room for Design to join Engineering and Product as a peer function.

I've seen this called 'EPD', 'PED' and 'PDE' but in any order it's the same: Three pillars holding up the customer experience. An EPD model isn't hard to do any more than it's hard for soccer players to stay in their lanes. Which is to say, it's really hard if you're new to it. When little kids play soccer they all just run right at the ball and try to kick it. The inside of immature startups looks a lot like that. Mature leaders move toward clear lanes that look something like the following:

Product — Identifying, de-risking, and preparing product opportunities for the company

Design — Designing *and scoping* implementations that integrate new opportunities with all historical decisions

Engineering — *Sequencing*, staffing, and implementing product designs in a way that continually accelerates future work

Another way to think of it is *the Cockpit Model*, where Product focuses not on directing activities but on providing clarifying information.

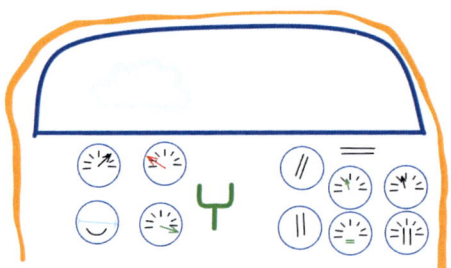

lets pretend this looks like a cockpit

In the Cockpit Model we have Product providing the instruments and windshield, Design providing the streamlined fuselage, and Engineering providing the thrust. Get these three pieces right and you have a craft any moderately skilled person can drive around. Without these three pieces in place even the most senior executive will struggle. The most skilled pilot cannot fly a broken plane.

If your company has a Product function that can provide rich product insight like a dashboard, a Design function that can make sure new changes maximally empower users, and an Engineering function that makes sure all the work gets done quickly, then you've built a powerful airplane. Deciding what to build and when becomes less of a struggle and more of an enjoyable, collegial conversation.

Engineering in the Value Chain

With that, look a little closer at the image from before:

We're looking for a way to measure Engineering's output. If Engineering's job is "sequence, staff & implement," how do you measure that?

We can't measure Engineering by whether our launches increase revenue because many other teams at the company contribute to that. And we don't want to measure something too narrow within Engineering like migrations completed or code quality because they don't represent output.

This gets easier if we rephrase the jobs of Product and Engineering with a higher ideal: **Product communicates the value** of each item while **Engineering communicates the cost.**

With that in mind, the relationships between functions get sharper.

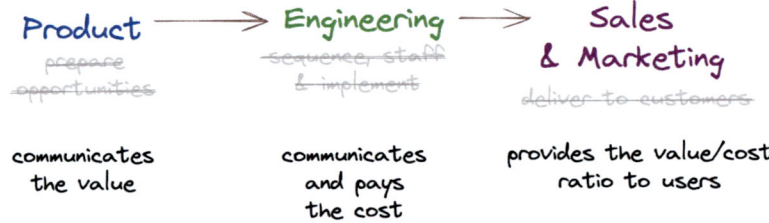

Engineering communicates cost (in time) because Engineering knows how to lower this cost. How to assign the right people to do the work, how to schedule it with other fixes and features, what level of future investing (or debt accumulation) is appropriate for the team implementing it, etc. Engineering leadership earns their keep by finding creative ways to minimize this cost while keeping engineers happy and growing.

Units of Engineering Cost

The engineering cost is the denominator for our metric.

The value, though. Ah, the value. That's what we want and it's the top of our metric. That's the numerator for our speedometer. The metric doesn't work if we only know the cost, we need Product to assign a *value*.

Whatever value the Product function believes a feature might have – that's how much credit Engineering gets for shipping that feature. It doesn't matter whether the feature actually has the supposed value — Product will try to improve their estimations over time — what matters is that if Product says it's worthwhile and Engineering delivers it then at least we know Engineering is working correctly.

Because that's where Engineering sits in the value chain: The point of Engineering is to deliver value as identified *and measured* by Product.

Units of Product Value

Actual value is defined in revenue and other top-line metrics, but *potential* product value — the amount of value that Product believes a future launch should have — is company-specific. Potential product value depends on what kind of things the company wants to get good at shipping.

For example, AWS has tracked shipped potential value by measuring the total count of press releases. They value an extremely broad set of mostly minor features that together form an unified ecosystem. Whether the feature gains traction or not, the AWS leaders gives themselves credit for a launch with each press release.

When Alyssa Henry, founder of Amazon S3, went to Square she developed a more sophisticated version of the same, but counting only major press releases.

When I took that from Square to Gusto, the Product teams there iterated further on this approach and increased the sophistication. We created a three-tier system that weighed major launches more heavily than minor ones, capturing both big features and small user-facing improvements.

In my last few roles I've used three categories of features: Competitive Advantages, Accelerants, and Table-Stakes with something like 30, 5, and 1 points each, respectively.

- **Competitive Advantages** are any features that let this company do something other companies can't
- **Accelerants** are any features that allow users to get more value from the product offering
- **Table-Stakes** are features that users probably assume exist either because competitors have them or because they're common patterns in modern software

The numbers assigned to each are arbitrary but they're still useful. Most companies have no existing baseline of shipped product value so a rough draft analysis goes a long way.

Over time your Product function will change how they evaluate features. There will be some politics and disagreement about how many points an initiative has but, importantly, that disagreement will be *within Product* and not something you have to worry about directly. As long as some point value is being assigned to prospective work and the CEO agrees that historical features with that same point value are about equally valuable then you're good.

Feel free to use the above 3 categories or any other system of measuring product value that fits your company, as long as it captures the *notability* of what could come out of engineering regardless of the final market value or the time it took to build.

Choosing a unit of product value

Product assigns the units of value by how much they value a launch. That allows us to align Engineering's success metric with its primary customer, Product. The more Product wants a feature, the more points Engineering gets for it, with no regard for its cost to implement. It's probably healthy for a company to periodically change its units of measurement here as it learns more about its own Product process.

There are some common failures for companies new to assigning product value:

- Assigning high value to something because it's very costly to build
- Assigning low value to features that competitors already have but are nonetheless valuable
- Assigning no value to some features because they're considered "must haves," as if our desperation factors into it somehow

It doesn't matter why Product wants the features or whether Product is excited about wanting or needing them.

I recommend, for your first pass, that you start with a very simple unit of potential product value: **"Features that would appear on a slide in an all-hands meeting."**

Make that the value metric until you come up with something that fits your company better. If it didn't or won't show up as a bullet point on an company-wide update, don't give yourself credit for launching.

Potential Product Shipping Chart

Once you have some unit of measurement you can create a historical baseline for yourself. Go back through past investor and board slides, press releases, launches, etc. and see what features you completed and when. Then plot them using a formula that you can use every quarter to hold Engineering accountable in a healthy way.

If you'd launched 9 major features in the last 8 quarters, the launch rate might look like this:

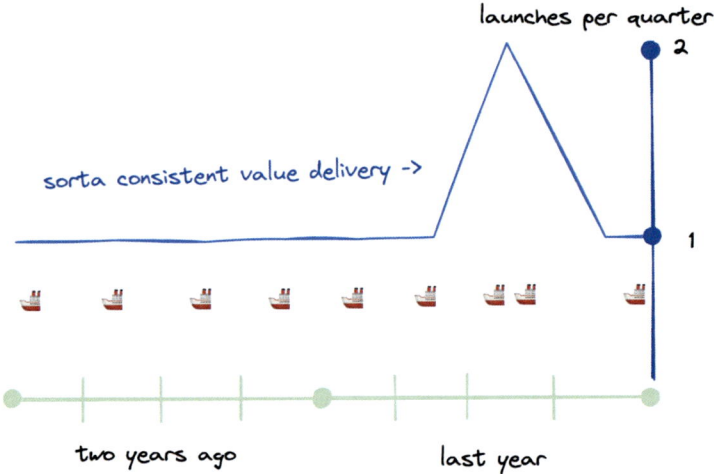

That gives you roughly the rate of launches but I recommend keeping a parallel metric that might be more useful: Divide those launches by the number of engineers at the company at the time. Hiring masks drag, and even if you can afford the additional headcount now you may not be able to later.

If you'd started with 8 engineers and added 2 people each quarter that same chart might look like this:

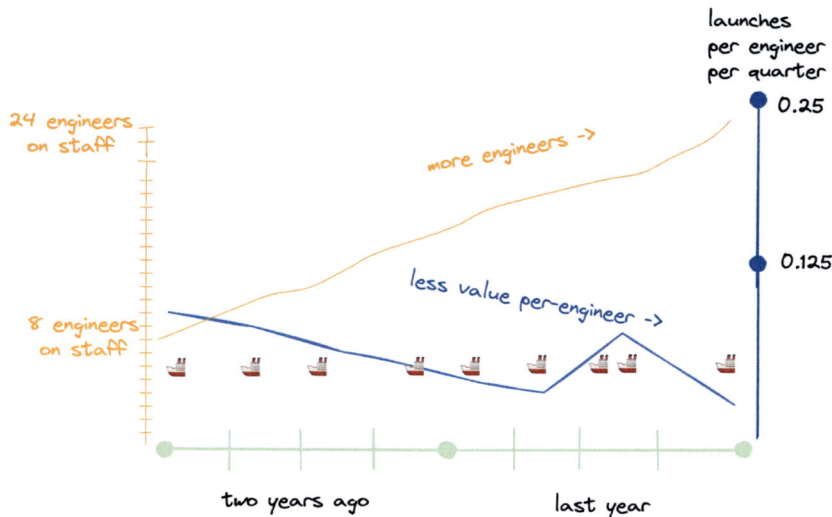

This chart of *Shipped Potential* is your new communication interface with leaders outside of Engineering. You can game, iterate on and scrutinize this metric without running afoul of Managing to Weird Shit and throwing the engineering work off track. It also helps your Product function be much more clear about the features they're prioritizing and why. Having to research and define the potential product value to an upcoming feature can sometimes reveal which features are actually worthwhile and which are merely somebody's pet project.

Regardless of your current role, if your current organization doesn't have something like a Shipped Potential chart you can make one yourself. If you want to make a point about just how much development work has slowed you can go one further and get historical headcount numbers. That'll let you start a conversation[11] around a chart that shows how many engineers it took to launch major historical features.

---

[11] When Alyssa Henry joined Square she found the historical numbers and sent an email to the whole company reflecting our current performance back to us. It was a useful wake-up call and totally changed how we planned features.

Using the Shipped Potential Chart

Engaging with Product

There are four ways I've used this chart and seen it used. The first way to use the chart is also a key part of building it: Engaging with Product colleagues to talk about the notability of work. The chart helps start a conversation around what they'd ideally like to do. If everything on our roadmap is 15 points what would a 100-point feature look like? What are the data and tools they'd need from Engineering to evaluate something larger?

This can be a particularly fruitful seed to plant in discussions about 5-year plans. With just a very slightly higher acceleration curve a 5-year timeline yields twice the development output.

However the conversation goes, when the Engineering and Product leads are aiming to maximize this chart they're perfectly aligned.

Engaging with CEO and Finance

The second way to use this chart is to engage the CEO and the Finance lead around when the company can afford to go slow if it means going faster later, and when it must be the reverse.

The tongue-in-cheek first answer will probably be "always go faster." Yet it shouldn't take long to differentiate between when fast would be nice and when fast is critical. There are likely upcoming funding milestones, competitive moats that need to be built quickly, or operating costs that need to be eliminated. If you're not yet ecologically dominant in your market then the competitive landscape will inform much of this.

Once these milestones are clear then you can back yourself into a hiring strategy. Hiring may be necessary to pay down some debts[12], but it also requires huge engineering time to recruit, interview, and onboard. So a slowdown like that should be a very carefully timed financial decision.

---

[12] You'll find a much fuller exploration of this dynamic in Part 4: Technical Debt Financing.

This kind of planning – like any investment strategy – requires understanding how heavily we want to go all-in on one opportunity at the cost of any future opportunities. This planning helps us avoid common mistakes like allocating 20% time to technical debt[13] or worrying about monoliths versus microservices.

Communicating expectations to eng leadership

The third way to use this chart is to communicate with your engineering leaders about what timeline you want them to optimize for.

You may ask your Growth Engineering lead to ignore this chart because they have other metrics you want them to maximize. Whereas you may need to drag your platform leaders into the technical weeds with you to work through which projects and migrations will have large impacts on this chart.

This chart may also spawn some essential conversations among eng leadership. The fastest way I know to get a rise out of senior, long-tenured engineers at a company is to ask them "what would have to be true for development to happen at 5x the pace?" They'll give you a long list of unstaffed and (possibly) important work.

Your own report card

The fourth and, frankly, most useful way to use this is just to hold yourself accountable. CTOs make countless little decisions in our work around staffing, culture, compensation, and process that all affect Engineering's effectiveness. This chart is ultimately our own personal report card.

Once you've made a Shipped Potential chart that shows the speed of Engineering you'll have an artifact you can share broadly outside of Engineering. Which may buy you a little breathing room to let your eng leadership do nuanced internal work that may only make sense to them and their engineers.

---

[13] This pattern is a sure sign of a leader who doesn't understand where technical debt comes from

Maximizing Output

When a 9-year-old runs a mile they start off with a sprint. Their little legs are flying and they're making great time. Then they walk the entire rest of the way.

When they're older they learn to pick a cadence that works for the entire duration of the distance they plan to run. And if they ever reach professional or Olympic levels they'll aim for the gold standard: A negative split. They'll try to run the second half of the course *faster* than the first half.

As an engineering executive your goal is that negative split. You want your teams to move briskly now, sure, but you know this journey is long. In fact, it never ends. You need your teams to ship early and often and to do it in a way that accelerates (even if slowly) their work over time.

The development pattern I see most of the time looks like a 9-year-old's sprint. It's a kind of Agile that celebrates sprint predictability (did we complete what we planned?) over real output (are we delivering value?) and makes sure engineers are only working off a queue of planned work.

The above is good but incomplete. It frames software engineering as primarily a management problem, one where supervisors plan the work of their subordinates.

No matter how perfectly we execute that process failure comes at us from two directions:

1. The 'subordinates' often know better than managers what a task really entails. And their knowledge is intuitive enough as to make a coherent articulation difficult. Task knowledge accumulates in the people doing the work.

2. The technical system itself has the power to direct engineer's labor with as much authority as management. Engineers are beholden to two bosses: An organization with business needs and a computer that stubbornly refuses to compromise.

The outcome of this common sprint-focused pattern is a hyperfocus on visible, impactful work and a high value placed on velocity. Which, counterintuitively, leads to lower velocity over time. Just like the 9-year-old running a mile, focusing on the immediate future leaves us unprepared for the rest.

The alternative is a planning model that values output over a longer time horizon. One where we ship immediately, for sure, but set 6, 12, and 18 month goals and make today's work set us up to reach those goals quickly.

We use Shipped Potential as our metric to guide us toward our negative split. With that as our north star we look at what actually speeds us up.

The Fundamental Forces of Engineering

Thrust and Drag

If it's 2018 and Softbank just landed an aircraft carrier of cash on your CEO then maximizing the output of engineering is easy: Just hire more engineers.

For the rest of us, our company is in an existential battle against both competitors and market irrelevance. We need to improve our output *per engineer*.

We maximize output by managing the fundamental forces of engineering.

The aviation metaphor continues to help us here, as the forces that push planes around the sky are fairly intuitive. Lift causes the craft to rise, gravity causes it to fall, thrust urges it forward while drag frustrates that forward motion.

We could put huge engines on an airplane that has poor aerodynamics and it would move through the sky, but at great cost. We could also make a streamlined body with weak engines and it would perform about the same. Engineering works along precisely the same principles: A large, senior engineering team working in a poorly-streamlined context makes no faster progress than a small, junior team facing lower drag.

Initiatives to increase velocity rarely involve these fundamental principles. Countless times I've seen or heard of a head of Product or Engineering who's determined to find some activity, some three-week initiative that can speed up the work. That's like a pilot trying to find better tailwinds at different elevations and getting creative with when to burn fuel. These are tips and tricks that marginally improve performance but this approach can never 10x the speed.

To actually improve engineering velocity we work to simultaneously increase the thrust of the engineering teams and lower the drag they face.

$$\text{Forward Motion} = \frac{\text{Thrust}}{\text{Drag}}$$

Thrust

Thrust comes from an engineering team correctly understanding a problem, knowing all the technologies necessary to solve it, and visualizing the path from start to finish. When a team has all three of those and no drag it's best to get out of the way because it's T minus ten seconds and they're about to launch your company into space.

As we work through the organizational models in this book you'll learn ways to place your engineering teams in precisely this position. We'll organize Engineering into several layers that give each team a clear idea of what needs to be done, we'll structure a team process that uses just the right balance of Agile for the specific team in their context, and we'll go deep into designing communication systems. All this together gives your teams the thrust necessary to get any job done.

We won't discuss how either effort or hours worked contribute to forward motion. The autonomous drive of an engineer is powerful enough that we don't need to waste our attention on perceived effort or total hours worked. The output amplification we're looking for is massive — doubling or tripling our engineers' output as often as we can — so it's a distraction (not to mention a real downer) to focus our attention on whether someone is sitting at a computer for long enough.

Drag

Drag is the giant lever where you as an executive need to focus your energy.

Drag is hard to see but easy to feel. It accumulates as the side-effect of all of our other work. When a company starts out — with that first line of code where development speed is infinite — there is no drag at all. Drag increases with each new source code file and each new dataset. Each new teammate adds to drag because now there's increased collaboration overhead.

As the system gets more complex it gets slower, drag increasing throughout until it can feel like every task involves bureaucracy.

This new complexity is essential; we can't win a market without adding new features and data and teammates. We executives, therefore, must structure the shape of this complex system so that the paths our people take through it are streamlined.

Our task then is the same as an aeronautics engineer designing a fuselage: An airplane will always face wind resistance but a good design minimizes drag while still letting the plane carry sufficient people and cargo. An airplane moves swiftly through the air only in the forward direction because the fuselage is designed for forward flight. Imagine if we put the engines on sideways and tried to fly it left. Many companies struggle through a technical and organizational context that feels about as effective as that because the work context hasn't been designed for the direction they're trying to go.

We don't have to eliminate all drag, we just need to lower the drag an engineer encounters as they do the work we expect them to do. That's a key distinction: Your technical stack sits on top of incomprehensible complexity including whole operating systems but you aren't asking your engineers to wade through the source code to Linux. The system you operate can be explosively complex as long as the cognitive load an individual faces in their tasks between here and your goals is minimal.

Drag reduction, therefore, is a matter of organization; of creating streamlined paths through the technology and the bureaucracy that allow Engineering to encounter minimal conceptual overhead while they work.

There are three major ways to maximize thrust while minimizing drag and they're the final 3 parts to this book:

- Through the streamlining structure of *Technical Coherence*
- Through careful Technical Debt Financing
- Through empowering engineering policies and culture

These levers are only directly available to the executive in charge of Engineering which is why this book is written for those who are in that role. If you're not currently in that role you'll need to apply these disciplines within your own area of influence and hopefully show enough forward motion that you can then spread your ideas across the whole org.

Part Two

Technical Coherence: A Theory of Engineering

In anything at all, perfection is finally attained not when there is no longer anything to add, but when there is no longer anything to take away, when a body has been stripped down to its nakedness.

It results from this that perfection of invention touches hands with absence of invention, as if that line which the human eye will follow with effortless delight were a line that had not been invented but simply discovered, had in the beginning been hidden by nature and in the end been found by the engineer.

There is an ancient myth about the image asleep in the block of marble until it is carefully disengaged by the sculptor. The sculptor must himself feel that he is not so much inventing or shaping the curve of breast or shoulder as delivering the image from its prison. In this spirit do engineers, physicists concerned with thermodynamics, and the swarm of preoccupied draughtsmen tackle their work.

In appearance, but only in appearance, they seem to be polishing surfaces and refining away angles, easing this joint or stabilizing that wing, rendering these parts invisible, so that in the end there is no longer a wing hooked to a framework but a form flawless in its perfection, completely disengaged from its matrix, a sort of spontaneous whole, its parts mysteriously fused together and resembling in their unity a poem.

— Antoine de Saint-Exupéry, The Airman's Odyssey

Technical Coherence

Technical Coherence is a three-step process to minimize drag:

1. We identify the **user experience domains** our company must support for the products we're trying to build

2. We identify the **shared product domains** that underpin multiple user experience domains

3. We staff **three layers of engineering teams**, the top two mapping to the layers we just identified and the third layer providing infrastructure.

This is an organizational design pattern that's exhaustive and can be mostly accomplished in a single meeting with your leaders.

One interconnected system

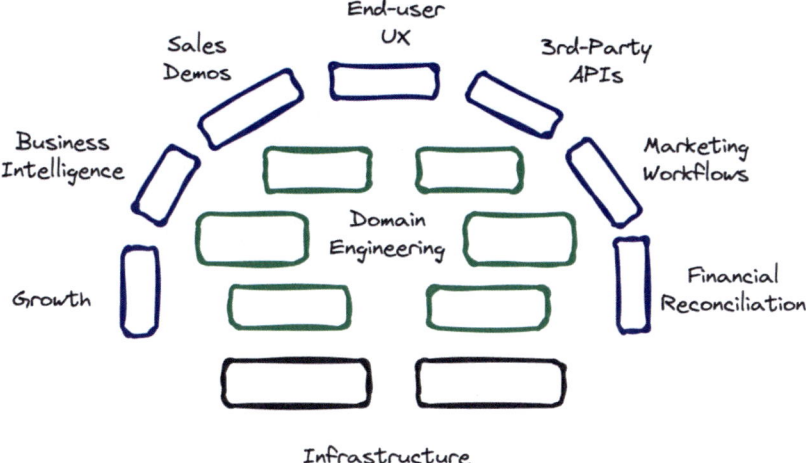

We create a low-drag system by first identifying all the users of the product, internal and external, and the interfaces they'll need, whether it's a UI or an API.

We commit to building these interfaces on one interconnected system, not addressing them piecemeal the moment an interface needs attention. These interfaces must be one system or else our users will have to sign in to many different sites, copy and paste their data around, etc. That would be a bad UX but it's also far slower to build and much harder to secure. So we need to think of it and all its dependencies as one interconnected system, no matter how broad and complex it is.

We use this surface area to develop the outer shape of our aircraft – finding precisely the most aerodynamic shape to get us from where we are to where we're going.

These interfaces must be owned and supported by Engineering. I consider this absolutely non-negotiable because if someone who is not an engineer builds one of these interfaces (via some low-code tool or a direct database connection) they can compromise the integrity of the system's data through writes and its security through reads. Many people can develop software but engineering involves certain guarantees of correctness – and if data is going into the system we'll need those guarantees of correctness. Also, to secure a system we must be able to enumerate all the things that a user might do with it. So Engineering must own all interfaces to the product, even if those interfaces aren't mentioned in the Product roadmap.

Underneath those surface interfaces we find layers of abstractions all the way down to some hardware running electrons through silicon somewhere in the world, both on users' devices and on our own servers.

The Shape of the Airplane

In 1935 Howard Hughes set the landplane speed record with his first aircraft, the H-1. He wasn't better than other racing pilots, he just streamlined the plane. His team made flush rivets – screws that didn't stick out from the body of the airplane – and carefully designed the engine cowling, hand-shaped the aluminum skin, and they added countless more little adjustments that minimized friction in the air.

They made these improvements before they ever flew it. They used the CalTech wind tunnel and the theories of aeronautics to design an optimized craft.

High-velocity software engineering doesn't have a wind tunnel and, worse, we don't really have a generalized theory of engineering performance. Lacking that theory we tend to focus on the process of flying the plane, whether it's plane-shaped or not. To drive teams faster through one-off projects, task forces, and innovative processes.

We might have a strong intuition for how much Infrastructure staffing is best but there's no predictive model we can use to test our beliefs. Finance and Sales can say precisely how much output comes from X units of input, why can't we?

Here's a situation you will almost certainly find yourself in, at some point: Your company has a Customer Support function that's growing in headcount much too fast. You want to add automation, tooling, and generally better leverage so you can decouple Customer Support hiring from customer growth.

Perhaps you'll want to staff one team for two years to improve Customer

Support tooling and APIs. Is that the right amount? Will that speed up or slow down other feature development? You might have the right intuition but how will you explain your headcount investment if your CEO or peer executives disagree with you?

The lack of a holistic theory of Engineering leaves us dependent on spending our social capital and pinning our decisions to comps from other companies. The lack of a theory also opens us to maybe adopting whichever new flavor of (cross-)functional teams or pods or squads or Agile is currently making the rounds.

We'd benefit from a way to think about all of the engineers, all of the technology, and all of the product together as a whole. Something at a higher level than activity and process that lets us see the entire territory all at once and plot our course through it nimbly.

There's just no escaping Conway's Law[14] so as we're creating teams and sub-teams we know we're actually crafting the technical system that will fly us toward our product goals. If we can't see the path ahead of time and streamline the plane for the journey then our teams are going to move slowly.

---

[14] "Organizations, who design systems, are constrained to produce designs which are copies of the communication structures of these organizations."

How To Design an Engineering Org

This book is my theory of engineering and *Technical Coherence* is its centerpiece: A framework for designing and staffing an engineering org from the outside in, starting with the shape of the plane.

It's a way to help answer some very hard questions:

- What proportion of engineers should work on infrastructure versus product?

- Should we pay engineers in infrastructure more? Less?

- Are there multiple hiring bars for teams working on different technologies?

- What's the relationship between frontend infrastructure teams and backend infrastructure teams?

- Is Data Engineering a part of Engineering? How about Data Science?

- What is the ideal relationship between Security Engineering and Product Engineering? And can we get away from the consulting relationship where Security always feels brought in too late?

- How much do we pay down technical debt and who does it and which debt?

We start from the outside in, working backward from financial realities to product needs and then into a technical structure. Just like Howard Hughes using the CalTech wind tunnel to test precise scale models of his plane, we first find the precise overall shape that our product must take *before* addressing the internals. We're all busy so this isn't a complex theory – Technical Coherence is a 3-step process that can be mostly accomplished in a single meeting with Product and Engineering leaders.

I call it "Technical Coherence" because it drives toward a single, articulable system that implements the full product, can fit on a whiteboard, and improves security, correctness, performance, resilience, maintainability, and development speed. It's a way to talk about all of our goals at once by focusing on how we'll build and staff a single system that would meet those goals.

This approach is the result of my own experience eyebrow-deep as both a staff engineer and a senior leader in the depths a dozen enormous SaaS product suites, some of which worked well but most of which were barely-contained garbage fires once they scaled. Instead of adding a clever abstraction on top of the work we identify what matters and remove absolutely everything else. Technical Coherence is the engineering leader's equivalent of revealing the figure trapped in the marble block and delivering it from its prison.

Fundamentally, only an executive in charge of Engineering can accomplish this process, hence the title of this book. It requires ownership of an entire system. Any sub-system is eventually going to collide with any other sub-system. When I was at Square we once thought we had two fully separate, non-overlapping engineering systems: Cash App and the rest of the company. This was true for years and the separation was stable. Then one day we needed to connect payments made in one system to payments made in the other. Suddenly we had one massive misaligned system.

That's not a rare story, either. If our Microsoft logins are used for not only Powerpoint but Excel then that's one unified system. Worse, that same login is used for XboX, meaning the leaders at Microsoft have been forced, in order to limit product friction, to find a way to treat their incalculably massive technology suite as one integrated whole across many different product lines.

What we're about to walk through draws on some of the insights from Domain Driven Design (influenced heavily by practical modern takes on DDD[15]), as well as Team Topologies, and some of the rare but very useful academic research[16] in software engineering development.

---

[15] https://learn.microsoft.com/en-us/previous-versions/msp-n-p/ee658117(v=pandp.10)?#domain-driven-design-architectural-style
[16] Bird, Nagappan, Devanbu, Gall, Murphy 2009 – DOI:10.1109/ICSE.2009.5070550

Infrastructure Gravity & Domain Engineering

You become responsible, forever, for what you have tamed.

– Antoine de Saint-Exupéry, in The Little Prince

Product vs Infrastructure

We're out to reduce drag and increase thrust by designing the right system for the job. That system will have three levels to the engineering org, mapping to three completely different kinds of work within the technical product stack.

Each company draws its own line between Product Engineering and Platform/Infrastructure/DevOps but the difference between them is clear: Product Engineering is "stuff the whole company wants" (making features) and the other is "stuff the engineers say we have to do, I guess."

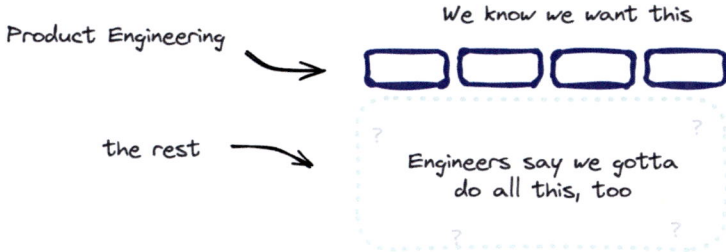

I've worked both above and below this divide and several times now I've led the entire span of engineering, from feature development down to whatever you call the bottom layer.

And I think I know why that bottom layer has so many[17] names: It's actually two separate things. Both are invisible outside Engineering but they require radically different architectures, leadership, and investment models.

The Product function peers down into the technical stack like a person in a boat, trying to see the depths. There's a limit to our perception from the surface. To see further we must plunge into the water ourself.

Everything visible when looking at the UX — the surface of the tech stack — is considered "Product Engineering." Below that the work is opaque and is assumed to be a cost center for the company.

---

[17] Infrastructure might be called 'Platform', 'SRE', 'DevOps', 'Cloud', 'Shared Services', 'Foundation', 'Systems', 'Core', or countless other names.

This view gives us two levels: The visible, and the invisible. And the line between them depends on the technical sophistication of the viewer.

This separation serves us poorly.

Partly because it prioritizes the merely visible over the important. To return to our aviation metaphor, prioritizing just visible product work is like making a plane out of just the visible parts: A fuselage, a flight stick, wheels, etc. We actually need a whole plane — with all the little details — if we want to fly.

The separation between visible and invisible work also obscures how the lower levels are more foundational in the system than the surface. If a feature breaks then just that feature is broken. If something deeper breaks then all features break.

I believe there are exactly three levels to Engineering in a product-shipping company, not two. These three levels exist no matter the size of the company and they do not necessarily map to the org chart.

They have many names but I call them *Product* Engineering, *Domain*[18] Engineering, and *Infrastructure* Engineering.

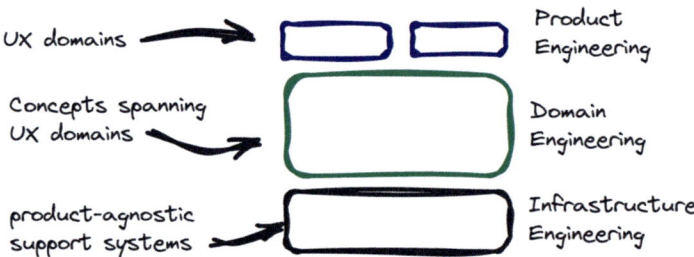

They all work toward different purposes, with very different constraints, and along completely different timelines.

Let me show you.

---

[18] Domain Engineering is a useful, existing industry term distinct from though overlapping with 'Platform Engineering'.

An Infrastructure Org is Born

There's a Big Bang moment at the start of every tech company. The first line of code is instant and the next lines are the beginning of a permanent deceleration in velocity.

The team focuses entirely on making features and the system gets more complex until, hopefully, it's actually useful. As new software engineers join the team they add further complexity and new patterns. Their excitement about features is tempered by a growing frustration with the underlying complexity. Folks tidy as they go but their goal is value creation, not cost reduction. The company hasn't yet earned the right to clean up its mess.

Then one day, usually when there's between 5 and 15 engineers at the company, one of them gets frustrated enough that they stop creating new features. They break away from the team and focus their attention on, say, the deployment script or the database config or the test suite.

Voilà. An infrastructure org is born.

This buys the company some time. With one person tending to complexity the rest of the team is free to keep piling new value into it. The infrastructure person fixes whatever problems exist, regardless of the type of problem. They can refactor source code, fix databases, debug BI dashboards, or improve deployments.

With their breadth of skills they're prioritizing across the whole company's needs, not limiting their work to just one layer in the technology stack or one app or technology. They can do this because they've helped create everything so far.

Other engineers may follow this person from feature development to infrastructure. The correct ratio between product work and infrastructure work depends entirely on the seniority of the product engineers and how much of the system they can reason about as they build new features. Anything the product engineers can't perceive or don't make time for the infrastructure team handles for them – so if your product engineers are junior you'll need more infrastructure people[19].

One day someone will be hired directly onto this infrastructure team. At this moment the engineering org undergoes a fundamental change and creates a problem it may not notice for years: There is now a person who's fixing a foundation without understanding what that foundation supports.

An external hire doesn't know the messiness of the product implementation. The features are, by definition, unique to the company so they can only learn them here. An external hire on an infrastructure team likely cannot upgrade, migrate, and reason about the system as a whole. They have other – very impressive – skills, but they can't improve product internals. To do that they'd first have to build product features, which takes them away from growing their low level skills that command higher salaries under titles like 'DevSecOps Engineer' or 'Cloud Architect' or 'SRE'. Going into product engineering would slow their career.

The product engineers may be justifiably impressed with these folks and the reverse is (hopefully) also true; neither group can do the other's job. One knows the specific product in detail, the other knows the patterns for software generally.

It's bad enough that the company may have just recreated the exact problem that DevOps intended to solve (one team writes software while the other one runs it), but there are well known solutions to that. The trap here is that, increasingly over time, some engineers are building features at the top edge of the system while others work at the extreme bottom of it.

Nobody is looking at the middle.

---

[19] This is why it's so unhelpful to use other company's Product/Infra staffing ratios. The sophistication of our Product Engineering teams (and the pace at which we drive their roadmaps) determines how much they leave undone underneath their features.

"Unique to this company, shared between features"

The middle of the system is messy.

Ask any engineer working at a large org what's the hardest part of making features and they'll likely say the dead center of the product suite is a giant mess.

Because 'middle of the system' is super vague let me be more clear. When the very first feature is created, all the technology at the company is in two categories:

1. The unique logic, designs, and choices that comprise that first feature

2. All the frameworks and tools to support changing and running that feature

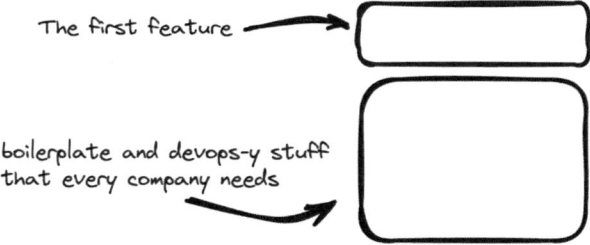

The top category is the implementation of the company's value proposition. Nothing up on top is particularly transferable between jobs because it's the company's competitive business advantage — it's by definition totally unique.

The lower category is the opposite – if you work here at one company you can get up to speed on it quickly at another. Every company must run unit tests and deploy software so these patterns don't change much, they're less a matter of invention and more of learning enormous vocabularies of implementations. You can get paid well for knowing that a process with an exit code of zero has succeeded and applying that knowledge across industry-standard tools.

If there's anything in this lower category that's a unique pattern — anything at all — it either should be deleted or it belongs up with the features.

As we add features the middle of the system reveals itself.

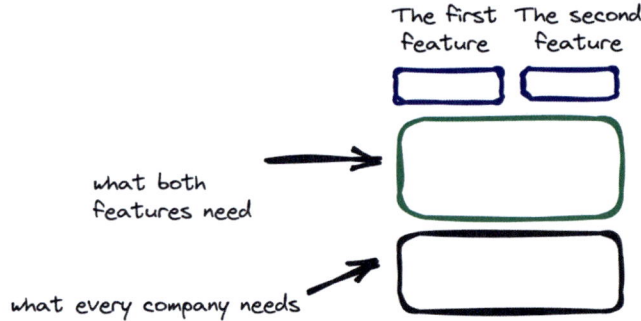

This middle layer is the glue that binds everything together. We can define it as "Everything that's unique to this company's features but not unique to any one feature." It's the internal semantics of the product suite as a whole, even if that product suite isn't a suite yet. Even if it's not a whole product yet.

Try to hire an infrastructure person from another company to fix your deployment and they'll have it done quickly.

Try to hire a software engineer to build new features and they'll do fine, as long as they passed the minimum technical bar and the feature is well specified.

Try to hire someone to work in this middle layer? They'll immediately stumble over the internals of your company's unique inventions.

How the middle atrophies

There's a good reason this middle is so often unowned or poorly owned. Two reasons, actually. I call them *Infrastructure Gravity* and *Feature Lift*.

They're hidden but powerful forces pulling our engineers to the extreme top and bottom of the technology stack, forcing individual engineers to perform heroics in order to keep the middle alive.

Infrastructure Gravity

This bottom layer has a gravity to it, pulling engineers lower into the tech stack and keeping them there.

There's a clear line in the tech stack above which most new infrastructure hires never venture. This line can be hard to see if you're looking at source code but once the code is running it's very clear: It's the process boundary for the running app. Anything that supports running processes on a computer is below the line. Your cloud architecture, config scripts, the test harness, etc. Every system call that a process makes is dealt with below this line.

Above the line is the internal state of the processes themselves (and, by extension, their source code). Not just the business logic of each application but also the software frameworks and libraries, the features' performance, and the semantic interconnections between all features and dependencies.

It's easy to see why this line gets brighter over time. Inside a running process is code written by, possibly, a new employee in a big hurry a long time ago. Imagine you have a choice to, on the one hand, understand and improve this code or, on the other hand, find a way to execute the process 5% more efficiently using transferable Unix and cloud skills. Which would you choose? Only the latter is guaranteed to even work and it's the one that provides the best career advancement in a role with the better pay.

This is **Infrastructure Gravity**.

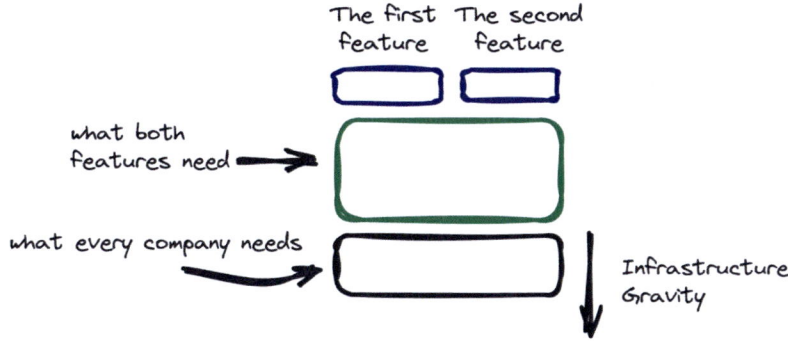

It pulls the people working on Infra / Devops / SRE / etc. down out of the middle layer towards the stuff below this line where they have a much higher chance of success and protection from the debt of the product work.

Luckily, a company needs this gravitational force to keep the most foundational parts of their system stable. Without it there's very little chance the system will be resilient or correct or secure. Infrastructure Gravity provides incredible value, it just also prevents people working at the bottom of the stack from helping out in the middle.

Since the people who work below this line can't easily see what's above it they have a highly localized knowledge of company problems. When folks down here feel inspired to invent something it will probably be an appliance that exists below this line. Perhaps a new deployment tool or an orchestrator or a scheduler. Perhaps a new message bus or an API that provides a novel datastorage pattern. These seem like big improvements if you spend all your time below the line (or currently work at Google in the year 2007) but the value they provide is minuscule compared to any improvement in this middle layer. What good is a new message bus if the user account data is broken?

Once you recognize Infrastructure Gravity in your org it becomes clear that staffing investment in Infrastructure Engineering is more or less a fixed amount. The minimum reasonable investment is also the maximum. Staffing this below the minimum might be catastrophic and adding more engineers here will not provide any increase in product value.

Feature Lift

At the same time there's an opposite force pressuring product-shipping engineers to stay at the top of the tech stack. **Feature Lift** pulls an engineer toward only the work necessary to launch a visible change.

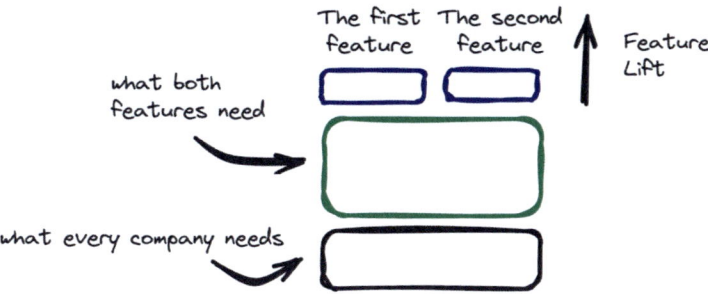

Some feature work can be accomplished either quickly or slowly, the quick way producing a hidden mess. An engineer working through a backlog of tickets will sometimes take the slower path in order to fix the system a little. But an engineer who consistently takes this slow path will be at odds with their team's mandate to ship features. Especially with teammates who don't agree about the necessity of going slow (or who can't perceive the mess). This leaves even the most well-intentioned and systems-minded engineer in a tension: Their ability to effect systemic improvements for the company requires them to spend social capital or to move to the lower, infrastructure layer.

This Feature Lift grows every time shipping fast is rewarded. Every time a product engineering team rewards itself for adding new value to users. Every launch, every performance review where someone has to write up their "impact", and every time a new shiny thing is shown to colleagues – the Feature Lift increases its pressure.

Feature Lift, like Infrastructure Gravity, is powerful and good. It focuses a team tightly on what matters most right now. Without Feature Lift a company wouldn't launch anything. There's an old phrase among SREs at Google about "running to stay in place" – refactoring and fixing the internals of a system in a cycle, forever, with no visible output. In fact, this is the most common reason

I've seen infrastructure tools companies fail: They're founded and led by people who hesitate to send debt-ridden, valuable features out the door (I've been guilty of this as a founder).

So this Feature Lift provides enormous value — it often provides a company's *only* value! But that value needs to be integrated with itself over time, the features consolidated with each other and the overall system smoothed out to make room for new features. No one feature is responsible for the company's competitive advantages. It's the system that holistically connects all features together, that integrates the features correctly, securely, and usably — that's what makes a modern product worth using and sometimes even worth paying for.

Creating features one after another without consolidating them is like making a linked list. Useful, to be sure, but the cost of traversing it is `O(n)`. If we were to structure these features in a better architecture then it's like storing elements in a binary tree. Which, under ideal conditions, can yield a far more efficient `O(log(n))` performance.

But a binary tree must be periodically rebalanced. If we just add items to it and never rebalance the tree it has the exact same poor performance as a linked list.

Once we recognize Feature Lift in our org it's clear that staffing investments here is roughly linear – each additional engineer can provide a marginal increase in feature development. Additional headcount here is like adding items to an unbalanced binary tree.

So we go to the middle to get *real* acceleration. The middle of the system is where we add more engineers and see a superlinear return on our investment.

The Gap – A space for Domain Engineering

As a CTO you own one large sociotechnical system of people and technology and you can reason about it as a whole. Systems theory gives us that a system has properties that none of its component parts have, so if we fail to zoom all the way out we will miss some quality of this big thing we're responsible for.

At this widest view, there are two ways the company is able to look at the technical system. These two perspectives map to Feature Lift and Infrastructure Gravity.

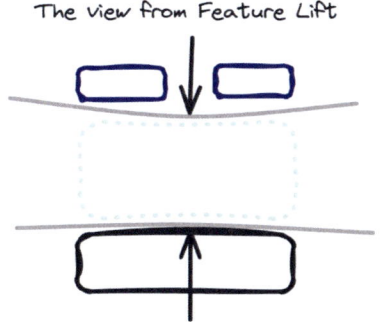

Most of the company perceives the system from the user experience. This view shows us all of the features pretty clearly. They're built for people to use (even the 3rd party APIs) so it's not too hard to see the system from this angle. This lets the company, and particularly the Product and Design functions, direct our work to improve the user experiences.

The other view is one that comes from the people who feel Infrastructure Gravity. These folks have developed sympathy for the runtime experience of the hardware and software itself. The memory use of processes, the latency of the network and filesystem writes, the data plumbing and storage patterns — this is a view of the system from the bottom.

There's no "Middle Layer Non-Gravity Non-Lift" pulling anyone to the middle.

The only thing that naturally draws engineers to look at the middle of their system is pure blinding rage. Given enough exposure to the neglected center someone will eventually make time to fix the things that bother them, whether they can make much progress or not.

Those heroes burn out quickly and they tend to be precisely the people you need mentoring junior folks, fixing security and performance, and doing interviews of senior talent. So lets not rely on only them to fix the middle of the system. These heroes feel responsible for the mess in the middle often because they helped make it but you, the executive, are actually the one person responsible for the mess. And that's good, because this is a point of enormous leverage.

As we look closer we can make better sense of what this middle layer actually is: This is *Domain Engineering;* your company's competitive advantage and your greatest asset as a technical leader.

Domain Engineering is the process of reusing domain knowledge to minimize the cost of developing products.

We see this in automobile manufacturing as shared chassis between models of cars. Or in software consulting as frameworks that generate software, making the work more configuration than coding. And at any company that would employ you or me Domain Engineering is the encapsulation and consolidation of the domain concepts underpinning more than one feature.

Let's make this concrete with an example.

Imagine your engineers are designing an authorization/authentication layer to let users access a suite of products. It might be tempting to say that, because every product depends on this, the whole user access system is Domain Engineering. Most of it is actually the lower Infrastructure layer because Infrastructure is what every company needs. And every company needs a way to store secure user credentials, perform authentication, handle password and token verification, revocation, etc. None of that supports the competitive advantage of the company.

It might be hard to imagine what would even exist in the middle.

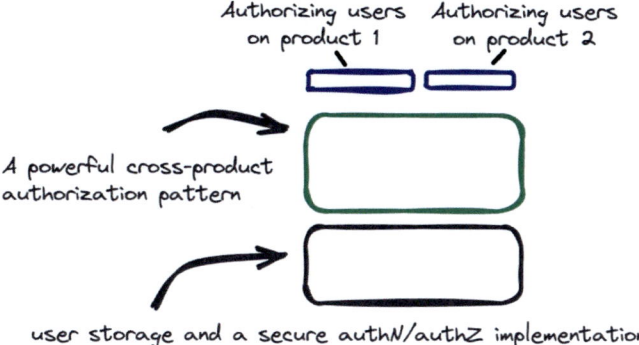

In between these two layers is a big opportunity to make investments in the company-specific way authorization and authentication work.

For example, what happens when a user is not allowed to see a resource, generally? Do they see an error page or get redirected somewhere? What messages are displayed and how is the user guided through a UX flow? How are resources partitioned between tiers of authentication sensitivity?

All of that should be solved in a library so an engineer can choose the right pattern when implementing a new feature. If any of those details require development time from a feature-shipping team then the product roadmap is being delayed unnecessarily.

Domain Engineering is the place to bake in all the company-specific decisions about authentication flows, service-to-service APIs, library integrations, error pages, and anything else that you don't want to drag on Product Engineering as they make new features.

And that's just the beginning. Domain Engineering is also the right conceptual home to shepherd better development of the company's competitive product advantages. It's easy to spot a competitive advantage — it's usually one of the oldest concepts around, it appears in virtually all of the products one way or another, and the implementation at some point becomes a huge mess. To GitHub this would be repos and commits, to Stripe and Square this would be the processes that create businesses and payments and purchased items, to Airbnb this would be the data and APIs that manage listings and reviews, etc.

Any competitive advantage at your company will be leveraged across many features and is therefore best owned by the layer of engineers underneath Product Engineering.

You might think this sounds like 'Platform Engineering' and it has a lot of overlap. I deliberately avoid that word because it's too easy to conflate that with "infrastructure" work and this layer needs to be insulated from Infrastructure Gravity. Anything pulling engineers down into infrastructure or up to building features will compromise Domain Engineering's success. I've seen Platform Engineering teams create novel data access patterns, implementing some custom CQRS or Event Streaming architecture that they think might be useful. That kind of work is not permitted in Domain Engineering because it doesn't improve the *conceptual domains* that undergird the product offering. It's arguably not even infrastructure – it's just more code and data.

Or, to think about this another way, how incredible is it that the competitive advantage of a company, something that appears in almost every feature, *wouldn't* have engineering teams dedicated to increasing the impact of its use across the company?

A Financial Model of Domain Engineering

As executives we're on the hook for the budget and outcomes from our org. It's not enough to have strong opinions on "right" ways to operate, we need to show results and we need accurate predictive models for those results. In the face of impending layoffs or fierce competition we need a financial model that lets us know what we can do to avoid failing our people.

My most useful staffing model is also my simplest. It starts with an assumption that product velocity at any given moment is a function of the development speed multiplied by current Product Engineering staffing.

$$Velocity(t) = Product\ Eng\ staffing \ast Development\ Speed$$

Product Engineering is the engine of our airplane and Domain Engineering has streamlined the fuselage to allow speed. Big engines are no match for having a fuselage in the shape of, say, a cube, so that Domain Engineering investment factors large into our velocity but it takes time to show gains.

Our thrust/drag ratio that leads to the development speed is a function of Domain Engineering staffing multiplied by how long Domain Engineering has been allowed to make progress.

$$Development\ Speed(t) = Domain\ Eng\ staffing \ast Time$$

Our Product Engineering staffing gives us linear results and our Domain Engineering staffing yields superlinear results.

This model lets us stop asking "how much headcount do we need for Engineering?" and ask a far more interesting question: "What's the timeframe in which we need to maximize overall product output?"

With both a constant (Product Engineering) and a coefficient (Domain Engineering) to play with we can articulate an investment curve that maximizes output at a time horizon we want.

Maybe we want immediate results for an important launch so we add headcount to Product Engineering and we have existing Domain Engineering teams join[20] Product Engineering on shipping some rapid features:

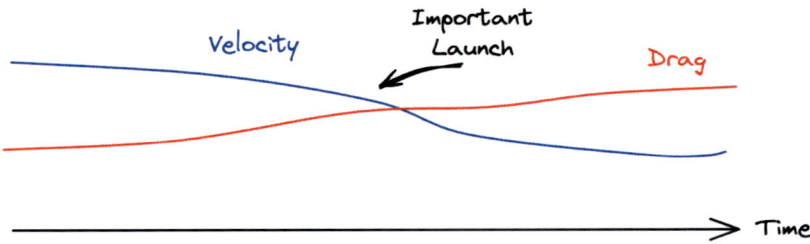

That gets us the launch we need at the cost of increased drag. We'll need to pay that down later if we want to speed up, but this is sometimes the right choice.

Or perhaps we don't have any immediate existential threat so we want to build up a head of steam. We can allocate headcount to Domain Engineering and train the senior engineers in Product Engineering to move them to this lower layer. That lets us lower drag until we have a system that's as flexible and fast as we need.

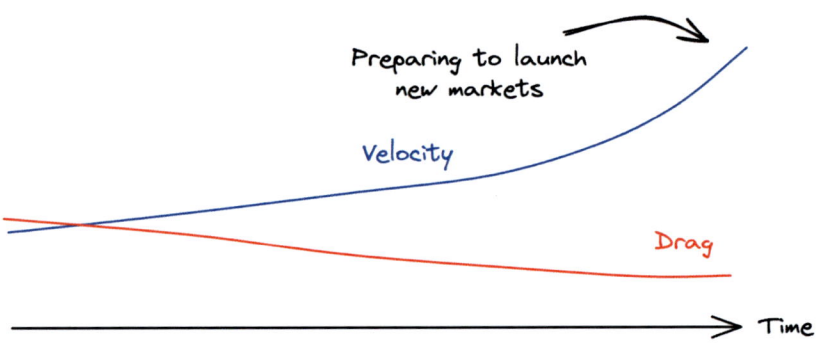

---

[20] If Domain Engineering teams can't ship features in a pinch then they're not Domain Engineering, these are Infrastructure teams in disguise

In either case we begin by picking a timeframe to maximize our output. We work backward from fundraising milestones, the competitive landscape, and our current market position into a staffing function that takes us where we want to go.

You may notice the lack of, say, units of measurement here. This is not an exact science but it's better than just piling more engineers into Product Engineering and Infrastructure, watching the former get ever slower and watching the latter contribute nothing to revenue.

Applying Domain Engineering with Technical Coherence

We're out to reduce drag and increase thrust by designing the right shape for our technical system and our people organization. Domain Engineering is a key part of this work but we need a way to apply it that is simple enough to communicate to all stakeholders and also useful enough that it'll succeed at accelerating development.

Technical Coherence achieves that with a 3-step process for organizing the problems Engineering faces and then creating a clear home for Domain Engineering.

Step 1: Identifying UX Domains

The first step to applying Technical Coherence is identifying UX domains. We'll be using the word 'domain' a lot so let's define it. There are many ways to use the term but we'll keep it simple: A domain is an encapsulation of something complex such that there's a lot happening on the inside with only a little on the surface.

Domain Encapsulation

An everyday example of domains are the specialized rooms in a house. Kitchens and bathrooms, specifically, are domains where a lot of related activities happen but there's not much connection between the rooms. All of our bathing happens in the bathroom. All our cooking in the kitchen, etc. Tina Fey once joked that her first New York apartment was so small she could stir a pot of spaghetti while sitting on the toilet. That's poor domain encapsulation.

$$\text{Domain Encapsulation Quality} = \frac{\text{\# of Related Concepts}}{\text{Inputs} + \text{Outputs}}$$

We could call just about any grouping of concepts in our system a domain but I recommend using *Domain Encapsulation Quality* as a mental framework for identifying domains. Which parts of the system have a lot of complexity inside them but could appear simple from the outside once the boundaries were clarified better?

UX domains are a particular kind of domain: They're all of the experiences a user might have with your product while they're in a particular role. Let's make this clearer by looking at an example of a UX domain: Onboarding.

A user signing up and figuring out your product is seeing many parts of your system but they all need to fit together and make sense. Later on you can show them advanced features but for now every part of the UX must gently hold their hand.

Executive Engineering

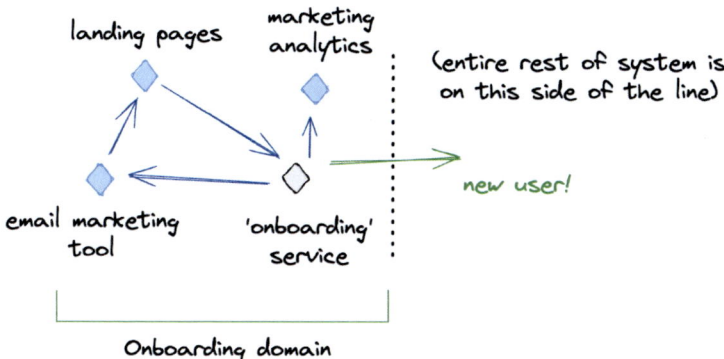

The domain encapsulation quality of this UX domain is quite high. The process of turning a new user into a fully-qualified participant in your product can be very complicated and is always undergoing experimentation but there's a single discrete outcome from it. At some point this domain produces a fully-qualified user that the rest of the system can use without ever having to know the details of the onboarding process.

The experiences this user will have after they onboard may feel like a new UX domain. It wouldn't be unusual for the post-onboarding experience to have different colors or UI patterns from what the user saw while onboarding.

As a user moves through stages of their lifecycle they may encounter many UX domains in your product. And your colleagues, partners, and other stakeholders may have UX domains of their own in the product that help them accomplish work.

The first step of Technical Coherence is simply mapping out these UX domains.

All of them.

The Experiences You Facilitate

Your company's whole suite of products — or single product, it doesn't matter — is simply this: The experiences you facilitate.

It might be hosting webinars, customizing AI models, or displaying a news feed. You might be Microsoft and it's everything from Office 365 to gaming hardware. Whatever experience you offer to any user is part of your product offering, whether you charge for it or not. Whether it's automated by computers or not. Whether the user is external or internal.

The field of UX research uses User Flow Diagrams to graph what we'll call UX domains and identify their dependencies. Folks who use Domain Driven Design would call UX domains an example of "bounded contexts."

To speak simply, let's use the term 'UX domain' to mean **all of the experiences you facilitate for a user while they're in the same role**. For example, if you're both a user and an administrator of an application the experience you have as a user is one UX domain while your experience as an administrator is another. They may look and feel the same but you wouldn't be too surprised if they weren't. Checking your email and writing a fresh new email are both in the same UX domain. Administering your organization's email policies would be a different UX domain, even if it's the same application and you're the same person.

We can break 'role' down into parts of the user lifecycle when that's helpful. The moment I sign up for a new email account I might be in a 'new user' role where the experience needs to be tuned for my unfamiliarity. There will be tooltips obscuring my view and tutorials popping up for me to opt out of. Later on I'll become more familiar and all the little tooltips and help text should get out of the way; I'm the same person doing the same thing but because I'm later in my own user lifecycle perhaps the software should consider me to be in a slightly different role.

The boundaries between UX domains are subjective but we only need to be directionally correct for our first pass of this work. A test for if two domains should be treated as one or separately is simply: Would the UX make more sense if we had one big team own all of this or two smaller teams?

You're responsible for the whole implementation

Whether your CEO knows it yet or not they will hold you accountable for providing every UX. This might be a robust and fast UX to a Customer Support or Operations team. And data to a Business Intelligence team. Even the most manual of processes will, once the headcount for that work gets too high, quickly fall to Engineering to automate so the company can achieve better margins.

So as you go listing out the experiences your product facilitates don't forget the manual ones. If a user calls a phone number and talks to somebody then it belongs in a UX domain somewhere. If a team at the company has to put packages in the mail for customers that goes in too. If somebody has to download a zip file of CSVs and FTP them to a bank, then that's yours too. If there's an experience your company provides to a user group then it goes in the diagram.

This might feel like you're taking on too much work for yourself by collecting all the jobs at the whole company and assigning them to yourself as one big engineering system. In practice this is accomplishing just the opposite: If you plan for everything to eventually be part of one big, well-organized system then you can avoid having to compromise your system in an emergency when it suddenly has to handle a new, foreseeable experience.

It's better to design a system that has room for every domain – even if you don't build it yet – than find there's no way to develop, say, a 3rd party API without a major overhaul of the internals.

Data Science is part of the product experience

Let's address a special case.

As I've worked with teams to draw diagrams of their product I notice that Data Science or Business Intelligence frequently get left out. Don't forget Data as a key user experience. It can be tempting to think of BI or Data Science as a function where brainiacs analyze the exhaust of whatever system Engineering built. Any company that thinks this way finds they can't get the right data out of Engineering and they can't easily use the insights from Data Science in the product.

What we call 'Data' is actually four fully separate things:

1. Coherent organization of source data
2. Informed transformation of data
3. Business Intelligence for human decision making
4. AI for automated decision making

Identify where, in your unique product, these four things need to happen. Every domain, at every layer will need a coherent organization of its data (#1) for both first-party use and for later analysis. So you can ignore this first bullet in your product diagram because it's mandatory to have everywhere — and should be a hard requirement in every engineering design.

But consider where the other three need need to show up as part of the product lifecycle and user experiences. When your exec peers or your board ask you about AI it'll be far more compelling to tell them where, how, and with what data you use it – along with a clear places where executives can plan changes – than some hand-wavy answer where it sounds like you're just sprinkling neural nets into your existing data pipelines.

Mapping out your UX Domains

We've defined UX domains as "all the experiences you facilitate for a user while they're in the same role."

Let's use this super high-level and broad definition to help us avoid overly detailed drawings. Consultants make money with impressively detailed UML diagrams but in a real team we need something that our colleagues can make immediate sense of.

To demonstrate how useful this is even at low fidelity, let's draw a complex product's UX domains using just five boxes.

Imagine you offer payroll services to small businesses in the US. This will involve some interface whereby a user can click a button called, perhaps, 'Run Payroll'. Before they can experience that perhaps their boss needs to go through onboarding to set up the company finances. And before that someone at the company has to go through a flow that converts them from a potential account to a real one.

In this example, you may have the following UX domains:

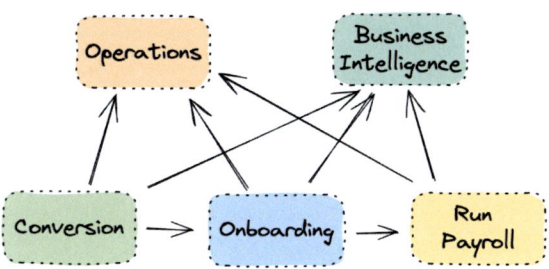

Each of these is a (large) set of experiences you facilitate for a user of some kind. Each one includes all of the experiences that a user has while they're in some role. Anyone who's accessing the Business Intelligence UX is a decision maker inside the company. Everything they need to observe trends and make decisions is a part of this UX.

There are clear dependencies between each of these UX domains – some of them rely on the data produced by others – so you can draw these as a directed graph, each experience producing data that is used by other experiences. These arrows are just a tool for us at step 1 of Technical Coherence – it can help us surface any UX domains we may not have thought of. Does someone in Customer Support need to call a user on the phone? Okay, well, "Customer Support" goes in the diagram and you can place an arrow to know precisely why.

In this payroll example there are both external and internal user experiences. I drew in only 'Business Intelligence' and 'Operations' as internal domains but Customer Support and Sales will need some kind of UX and there may be some kind of financial reconciliation team that needs a way to write data to the product somehow.

We start developing Technical Coherence by drawing these UX domains. All of them. If it's an experience that we must offer one way or another then it's part of the product. The reason we do this is because the product has likely been built as if there were just one or two first-class experiences, with others bolted on later. Just because the Finance or Customer Support UX isn't as urgent doesn't mean it's not as important. Your airplane may only need the landing gear to work later but that doesn't make this part optional. Failing to calculate the financials correctly or failing to support users is as much a risk to the company as an underdeveloped product.

So if your CFO requires some way to download CSVs with correct financial data from the product then that is a non-optional UX domain. Put it in the diagram.

If Customer Support needs a way to reset a user's password or access user data then Customer Support's experience is a non-optional domain. Put it in the diagram.

External partners need a way to retrieve their data? The Marketing team needs a way to see user outcomes based on what ad campaign first brought them in? The Security team needs to monitor an audit log of internal actions? Into the diagram they go – grouped into whatever experiences an individual might have while in the same role.

To actually write out the UX domains we need to know what our whole product is. What's the set of experiences we actually provide and to whom?

Getting to this high level dependency graph can be pretty tricky but not for any technical reason. You and your CEO and your Product leadership have to align on what the company actually does. As of this writing I've twice been in the situation where every executive agreed that "the product" was what end users paid for, but they simultaneously needed many other interfaces from Engineering: administrative dashboards, control panels for 3rd party users, data products for existing users and separate data products for prospective users, BI interfaces that allowed rapid trend analysis, and an integrated customer support system.

My job in the above situations – and likely your job at some point – is to demonstrate that all of these interfaces, these experiences, are surface area of just one integral system. And that the system can service any number of experiences very cheaply as long as it's built with those experiences in mind

You may find, as you've read this, that Technical Coherence sounds like an expensive and slow, bureaucratic way to develop a product. Like something from the pre-Agile days. I sympathize with that feeling but this up-front work is just the opposite: In a resource-constrained environment we only have a few shots on goal. Maybe just one. We may not have the time in between rounds of funding to rework a system that was under-designed. So as we sprint toward the MVP for our immediate goals we must set ourselves up for a smooth transition to the next goals at minimal cost.

Therefore, once we have *all* of the UX domains we're finished with the hard part. And we're finished with the first of three steps of Technical Coherence alignment.

Step 2: Identify Shared Domains

The next step is to identify the domains in the product that are shared *between UX domains*.

These shared domains are the conceptual areas that comprise the whole product suite. As we list them we see these are places where hard problems are encapsulated away, where proprietary business logic transforms data for a purpose, and places where the competitive advantages of the company are encoded.

More simply, they're also nouns and verbs that appear in documents and whiteboards when your company talks about what the product does. You might find yourself writing "The users haven't been responding to push notifications." If your product has much logic around push notifications then it might have a high domain encapsulation value and be worth managing as a domain.

The competitive advantage inside the product

I find it easy to identifying shared domains because they're the things engineering talks about the most. Let's find a couple examples from the payroll company we've been drawing.

We want to identify which domains – *not* datasets, datasets have zero domain encapsulation — might exist underneath the blocks in the payroll company's UX domain diagram. We're looking for conceptual areas where there are limited inputs or outputs with a lot of complexity inside.

Like bank integrations.

Somewhere in a payroll system there has to be an encapsulation of the actual movement of money. This kind of work requires installing into the system an intimate knowledge of banking APIs, timing, status codes, encryption schemes, audit logs — all in a financially compliant way.

Or messaging to users.

The conversion UX domain likely needs to confirm contact information for a user at some point. Operations may also need to contact users via email or text and there's probably some automated email or text message that happens once money moves correctly. This is at least three ways messaging to users happens for three different reasons. We can have each product team implement the bare minimum messaging for their individual features but that's going to result in slow development and a buggy, inconsistent messaging pattern with more overall code size than if we just did it correctly, centrally, once.

So let's draw our domain chart again. This time we'll leave out the dependency arrows (they were only there to help identify UX domains) and draw the UX domains as if they're the surface of a deeper system. Let's see how these two domains inside the product ('Bank Integrations' and 'User Messaging') connect to the surface area of the system.

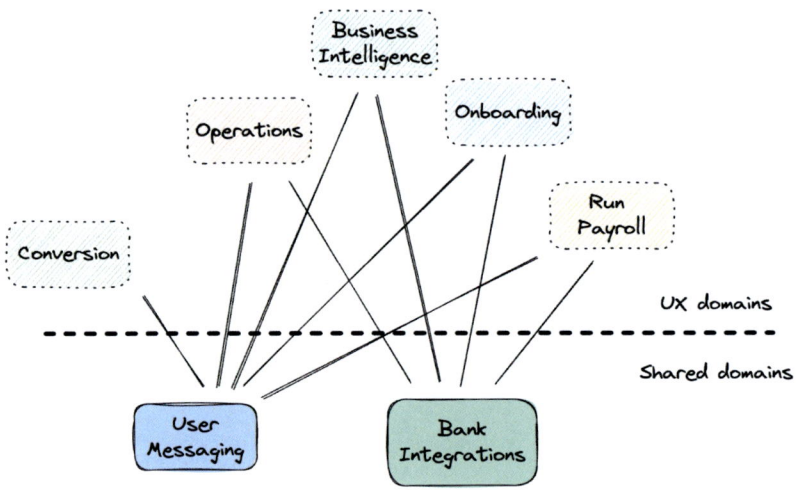

User messaging happens at various parts of the user lifecycle, triggered by various actions like the user signing up, someone in the Operations dashboard writing to the user, and by automated payroll actions.

Bank integrations get set up during Onboarding, used during Run Payroll, and managed by Operations.

And all of it needs to be visible in the BI interfaces.

As we identify more shared domains and add them to the image the lines connecting UX domains to shared domains become impossibly messy — just a solid sheet of ink. A pattern emerges: The UX domains, as a set, depend on the shared domains, as a set.

This separation between UX domains and shared domains is inescapable — we can either embrace it or waste a lot of time. If we try to fit our product development work into the framework where everything relates to a user-facing feature we'll get one of two things: 1) disjointed, redundant, minimal implementations of banking and messaging built in a hurry by product engineers or 2) burned out martyr engineers who made these shared pieces resilient but have lost trust that leadership can build a stable product.

The distinction between these two groups of domains is also critical to our planning. Any one of the UX domains could completely change or even disappear. We might, say, pivot away from B2C and instead get our customers through a business development channel. In that case we'd swap out the entirety of the Conversion domain with a completely new one. But we'd still need messaging nearly as much as we did before, and we don't want all of the transactional and other messaging to be ruined by the removal of some UX domain.

Put another way, UX domains can be staffed and developed at a level appropriate for how much we value those users and that experience. If, say, Business Intelligence isn't critical right now we can put minimal staffing on it. But the shared domains support the entire product suite and must be well-built or else they're a drag on both new feature development and system-wide stability.

So let's remove all those lines and just assume that roughly all the UX domains depend on roughly all the shared domains.

We fill in a couple more shared domains and we get a better look.

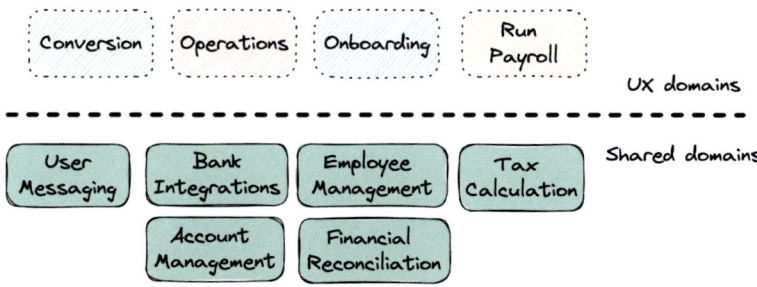

It might look at first like we've separated frontend and backend services, but that's not quite right. We're looking at the semantic dependencies of our product, not at the files and directories and services that might implement it.

Each UX domain may require browser or mobile UIs, backend services, and any amount of other APIs just to serve its users. And the shared domains may be implemented in entirely backend services but more likely they have pieces in every UI.

We've separated our system along two dimensions. The top and bottom are split so Product and Design can work on the top, in collaboration with Engineering. This top — the UX domains — can be planned with a traditional product roadmap. The bottom then are solely Engineering's responsibility and require a different investment model along much longer timelines.

And along the other dimension we've separated each domain from its neighbor. This might seem obvious or trivial but I've personally seen multiple implementations of a payroll product where bank integrations were not separate from the rest of the system. When we talk about the shared domains as separate concepts we give ourselves the ability to staff them, organize them, and implement them as separate concepts.

This separation is a huge drag-reduction tool. An engineer's forward progress is limited by how many concepts they need to work with between the start and end of a project. When we identify the shared domains we encourage encapsulation which allows engineers to move rapidly in a simpler environment.

Most of the companies I've worked at or advised have fallen into the trap of putting more energy into the UX domains at the surface than into the shared domains underneath them. This is a mistake: We're only as strong as our weakest dependency. The companies that I see fully staffing both layers find their product velocity increasing consistently. Partly because of the drag reduction from domain thinking generally, but also because shared domains of the product are a key part of the company's competitive advantage. The easier it is to build on a competitive advantage the more advantage it yields.

Once we've enumerated all of the shared domains we're now done with the second of the three steps of Technical Coherence alignment. Everything so far you could do with some engineers, product leads, and designers in a single meeting. This next part is where we go deep into how, specifically, we'll drive accelerations and improvements through just the engineering parts of the organization.

Step 3: Staffing the Breadth and Depth of Engineering

The 3 layers of engineering are distinct in how they work, how we staff them, who they serve, and how we incentivize the engineers.

Product Engineering

Product Engineering **creates product features.**

Product Engineering is what most people in the company think Engineering does. It's creating and improving any of the various user experiences that the company deliberately offers. Whether these are external user experiences or internal user experiences for employees of the company to manage the system, if someone needs to use part of the product then Product Engineering enables that.

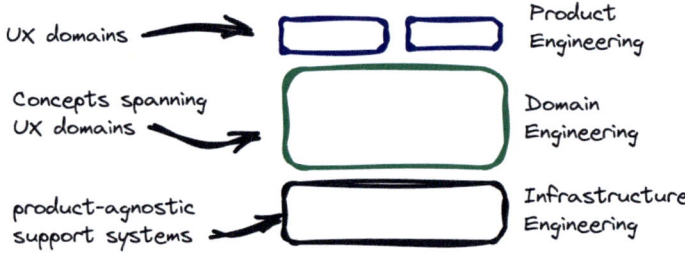

Domain Engineering

Domain Engineering **owns what is unique to this company but not to any one UX domain.** This is how Technical Coherence delivers its thrust.

This work is often unstaffed yet it's extremely valuable. Senior people in both Product and Engineering yearn for more staffing here. The engineers want it because they know it improves the work of every engineer. The product leaders want it because they know it unlocks new roadmap possibilities.

Some of the advantages of Domain Engineering are obvious but the most surprising one is how it delivers value to Product leadership. **Domain Engineering primarily serves the Product function** of the company, acting as the infrastructure for Product the same way Infrastructure Engineering delivers foundational improvements for engineers.

CPOs create product roadmaps based on what they believe Engineering can deliver. They look backward at the previous year's launches and then assume there's about that much capacity going forward. If the CPO wants to pot more interesting stuff on the roadmap they talk to Domain Engineering teams about unlocking new capabilities. Perhaps they want to roadmap new banking integrations so they work with Domain Engineering to encapsulate and augment the financial primitives in the product. Or they want to spend the next year finding multiple new revenue streams so Domain Engineering spends the current year abstracting away the things that make feature development slow.

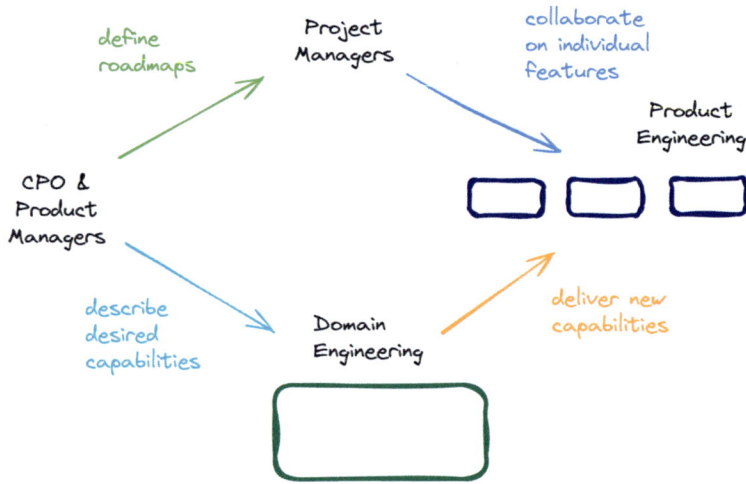

Product Engineering delivers on our short-term goals while Domain Engineering lays the foundation for longer ones. Domain Engineering allows Product leaders to dream about and then realize the brightest possible future for the company.

This is a key difference from how Platform teams might operate. Often we try to keep our Platform investments to just about the minimum necessary because that work is a cost center that competes for resources with product development. By separating Domain Engineering from Infrastructure more clearly we can plan these two layers of engineering work appropriately.

Throughout this book you might notice a tension I draw between Engineering and Product. That tension is real and the companies I've seen that lack that tension have either a disempowered Product function (acting as project management) or a disempowered Engineering function (acting as a consultancy). So the tension is inherent but what we *do* with the tension is up to us.

The most tragic failure mode I've seen — and it's really common — between two strong leaders of Engineering and Product is that each leader does what they alone believe is right and leaves the other to scrape for resources and to participate in internal company politics to advocate for their unmet needs.

This puts two extremely valuable disciplines at odds with each other.

Domain Engineering encodes into the org chart a belief that both Product and Engineering should be empowered. Product deserves to have a (large) team of engineers allocated to refining and improving the foundation to increase the scope of what Product can achieve. And Engineering deserves to have a (large) team of engineers allocated to ensuring the continued maintainability, correctness and uniformity of the underlying product ecosystem as feature value gets piled onto it. Domain Engineering does both with just one set of engineers, and precisely the engineers who are most driven and passionate about doing it.

Domain Engineering work is to Engineering what the flush rivets were to Howard Hughes' H-1 racer. What the streamlined body of a hull is to a record-setting aircraft. It's a way to ensure all of the work from the very top of the technology stack to the very body is not just streamlined in some generic way (that would be a sphere), but streamlined to fly straight forward *toward our specific product goals*.

My favorite outcome of Domain Engineering is how it aligns Product and Engineering leaders. Instead of conflict over short-term versus long-term goals both functions get to plan a steady increase in the development speed of features forever, continually. It requires foresight because it means the company can't allocate 100% of engineers to feature development now and forever, but the Shipped Potential chart reveals pretty quickly how slow it is to have everyone building features.

Scaling this up

As your organization grows you'll see individual products start to split along these Product/Domain Engineering lines. A large enough Product Engineering team will need it's own layer of Domain Engineering for each group of domains. Our teams and technology are fractal in that way — the conceptual patterns that work with twenty engineers scale to many hundreds.

Infrastructure Engineering

Infrastructure Engineering provides **what any company would need.**

This is messaging systems, datastores, repository strategies, testing suites, the CI/CD flow, observability tools, a runtime environment, etc. Infrastructure Engineering delivers the large mass of useful tools that any modern company in the same industry would use. The technology at this layer must be absolutely product agnostic. Just like a networking stack or a file system is not written with knowledge of the specific contents that will be placed into it, it's important that the technology at this layer be fairly generic. At least until the product gets so massive in scale that the tools need to be tuned in particular ways — but at that point that tuning is actually part of your competitive advantage and now may belong within Domain Engineering.

This system needs careful technical design because it can either lower drag significantly or massively increase it. Infrastructure Engineering teams must offer engineers one pattern for each technical problem. One way to write and run software, to store and processes data of each type, and to test code changes.

At many companies one may encounter, say, a monolithic app alongside a monorepo that contains services alongside multiple repos each loosely connected via heterogenous service architecture. Each of these is a fine pattern —no one better or worse than the other except for how well supported it is — but using multiple approaches is a terrible pattern. At Square we had precisely all three of these patterns and for a while we kept track of when new hires would try to solve the mess by introducing a fourth.

A company allowing a proliferation of duplicate patterns soon finds itself afoul of the most true principle of development tooling:

> "Any development pattern is better than two development patterns." — Jack Danger, just now

This will require intense empathy with Product Engineering. Infrastructure engineers have big opinions on how to write software but, thanks to infrastructure gravity, they typically aren't informed about how to write product software at their current company. And yet they need to provide one pattern for each problem type, ensure Product Engineers love it, and avoid adding a second.

The best teams working at the Infrastructure layer are laser-focused on the real needs (current and future) of the actual engineers who use the system. It's too easy to invent abstractions here or to continually solve one's own problems — or one's past problems from a previous company.

The Canvas

The UX domains give us the breadth of the engineering teams – the wide set of interfaces they must support – and the layers of work give us the depth. That's a two-dimensional surface across which we arrange our people and we must be thoughtful about who we put where and how heavily we staff each part.

There's a specific staffing process I'll recommend in a moment but first let's look closer at the dynamic that introduces this multi-dimensional canvas to begin with.

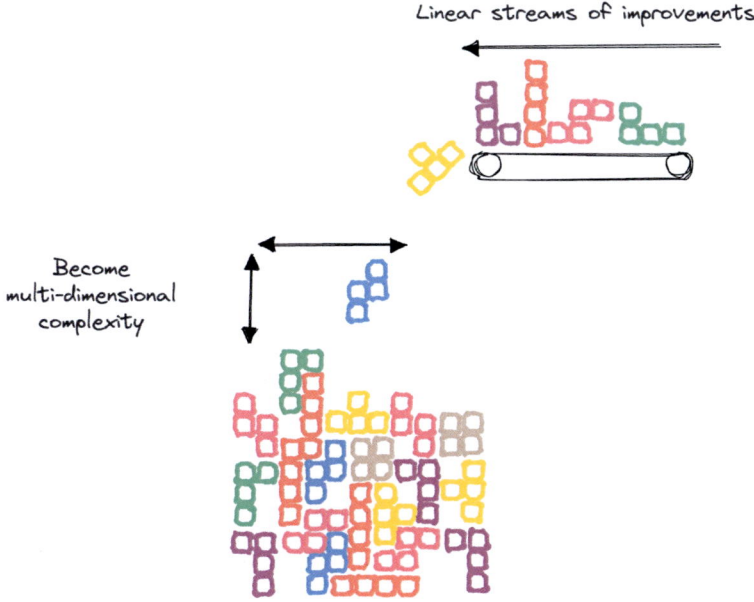

Improvements to the technical system happen linearly, though concurrently, across many engineers. Those improvements accumulate into a mass that has many dimensions and tends to get larger across each dimension over time.

We value two things about the stream of improvements to maximize our Shipped Potential:

1. How fast we can make improvements

2. How much product value each improvement contains

What we value about the system they become is almost entirely the same, but with one additional and important quality.

1. How fast we can make improvements

2. **How correct and resilient the system is**

3. How much product value each improvement contains

Stakeholders will rarely ask for systemic resilience. Partly because that's a mouthful but also it should be safe to assume a professional won't let something go haywire. When a colleague asks Engineering for a feature they shouldn't need to specify that existing features must also keep working. Therefore the correctness of the system is Engineering's responsibility, despite it never appearing in feature requests.

Our staffing needs to reflect these three systemic properties, *in this order*.

First, we need to be able to make rapid changes. Without this we can't quickly fix any problems with the other two.

Second, we must be able to assert that the system is working correctly at every level. Resilience isn't just about uptime and data integrity. The best security engineers I ever worked with were at Square and they taught me that security, simplicity, resilience, maintainability, performance, correctness and development velocity are all neighbors. When a system is far from one of these properties it's far from all of them. If something is hard to maintain then you can't patch it when there's a security vulnerability, add new features, or make it faster. If something is working incorrectly it's because there's something happening we didn't expect, which means our development process didn't reveal the consequences of a change and there may be security implications.

We aim for a resilient, correct, simple system because everything that comes with that is (usually) what our stakeholders need even more than they need new product value.

Lastly, we must ship high product value with each change. This comes from increasing our system's expressiveness (the ability to make a new feature with the least code) and consolidated product feature dependencies (the work that Domain Engineering does).

With an understanding of the two-dimensional space of the work and a clear goal of these three systemic properties, here's how I approach staffing the full breadth and depth of Engineering.

How to staff each UX domain and each layer

There are times of crisis when everyone in the company, all of Engineering and everybody else too, need to sprint to launch something new or fix something broken. At times like this the staffing is really simple: Assign everyone to work on the critical product areas, allocated by how bad the crisis is in each area. Despite all the abstract nuance in this book, when the company may not be viable you'll see me in the trenches with everyone else sprinting straight for whatever metric improves our financial projections.

What follows is an approach I recommend when your company is *not* in a crisis, when you're optimizing for the long-term success of the company, overtaking competitors and increasing competitive advantages, rather than sprinting to a short-term win and then falling over exhausted.

I know of many companies that use the crisis-mode strategy during stable times. Even with many years of runway some leaders staff product initiatives that maximize goals only weeks or months away, ignoring further timelines. I recommend having goals across short-, medium-, and long-term timelines concurrently if you want your Shipped Potential chart to have an upward curve.

First we allocate the minimum number of the most skilled Infrastructure people to the Infrastructure layer. This layer, riddled with Infrastructure

Gravity, requires some experience with systems. You can't afford to let someone invent commodities like a service mesh or generalized persistence abstractions — those problems have all been solved years ago. The minimum staffing here is also roughly the maximum: You want this staffed enough that the team can succeed and continually rebase the company onto the current year's best technologies. Any more than that is a costly distraction.

Next, we add as many engineers as possible to the Domain Engineering layer. There aren't that many people in Engineering who *can* work at this layer and who are also interested in it. This layer is near-impossible to overstaff because as these teams make improvements even your most junior new hires at the top Product Engineering layer become empowered. And the people qualified to work on Domain Engineering tend to enjoy switching back to creating features once the mess is cleaned up, so the risk of starving yourself of feature development is low.

Everyone else should be assigned to an appropriate Product Engineering team supporting a specific UX domain. Staffing domains — both UX domains and shared domains — is somewhat about interest and context so you'll see individuals fit better in some domains than others.

Over time the line between Domain Engineering and Product Engineering will prove porous, as the work sometimes moves from one layer to another. The more senior people on Domain Engineering will eventually pay down enough debts, clean up enough messes, and create enough internal platform semantics that they become free to work on features – making use of the improvements they just finished. And the more senior people on Product Engineering will graduate from building features to cleaning up their own — and their colleagues' — messes. Not only are these senior Product Engineers often the most effective people to clean up those specific problems but this is a critical maturation stage in an engineer's career and should be encouraged.

You might have a small eng org and too many UX domains to assign UX domains 1-to-1 to Product Engineering teams. Each Product Engineering team can own multiple UX domains but you'll want to consolidate the UX domains so that each team can focus on the same user lifecycle. This gives the engineers a chance to develop user empathy over time, a powerful tool in their ability to ship value faster.

I make it sound like you'll have a large pool of unstaffed engineers and a totally blank slate. Invariably, you'll apply this kind of plan to an existing org that already has teams, products, and a tightly-woven organization of one shape or another. I recommend the above staffing plan as a north star that you can use as you make incremental adjustments to an existing organization. The above process is the weights and biases you can use as you iterate toward an ideal.

Accepting the inevitable

At small companies it may might seem wasteful to staff all these UX domains and layers. Unfortunately, this work is not optional. Every experience the product facilitates will need to be supported, somehow. And the engineers will need to work at all three layers or else the code simply will not work and major, surprising, crippling debt can appear in your system.

We don't choose whether the work happens but we do choose how it happens and whether it's efficient.

If you run your software literally anywhere you will be using hardware and you will have some way to deploy your work to it. That's the **Infrastructure** layer. And when you build your second feature you can either copy-and-paste parts of your first feature into it or you can start doing the work of **Domain Engineering**. So even if it's just you running your own solo business these three layers are unavoidable.

Fast tests, continuous deployments, and a well-factored system are efficient at all stages of a company's life. Even in the early days a code change can be live in twenty minutes.[21] If it takes any longer — or requires any manual steps — then there's been a debt introduced somewhere that is dragging on your engineering speed. I know of no company so well capitalized that they can afford that for long.

---

[21] I've yet to see a deterministic test suite that can't be made to run within five minutes if pressed. Ten minutes is therefore an acceptable window. The same goes for deploying – five minutes is possible if the deployment stack is thoughtfully designed and ten minutes is a reasonable upper bound.
These numbers hold true for tiny greenfield apps as well as ten-million-line monoliths, as long as the tests are parallelized.

Applying the 3 steps of Technical Coherence

Shaping the hull

Technical Coherence, like the drawings in this book, is actually pretty simple. The first two steps happen in a single meeting with your most senior members of EPD.

Product and Design take the lead on step 1 to outline the UX domains. The goal is to make a comprehensive map of all the surface area of the system. The meeting attendees should identify the points of articulation and dependencies between each UX domain whenever possible, helping guide their thinking about user flows through the product lifecycles. It's important not to think about the internals of the domain because that will quickly waste time. No matter how complex the UX domain is if it's clearly one domain just move on, we're only listing them here — we empower the teams who own the domains to figure out the internals.

Then Engineering takes the lead identifying the product-internal domains which are dependencies of multiple UX domains. Step 2 here typically takes longer because the team might accidentally name datasets as domains and have to figure out the difference. A domain is an encapsulation of complexity, so there need to be some kinds of operations that happen within it. At this step we're not describing our current implementation, we're answering a more interesting question: If a competitor implemented our product what are the internal parts that would absolutely *have to* end up the same? I've had luck time-boxing this process to about 15 minutes and letting engineers work out the details afterward – we only need to be directionally correct in this meeting.

From there you can proceed to step 3 and actually think about what organizational structure you'll set as your new north star. The drawings from steps 1 and 2 should become more or less the staffing plan for the top two layers of your org. Staffing Infrastructure will depend on the maturity of the current infrastructure and how much standardization, security, and performance you need to empower your higher two layers of engineers.

It's at this point that you work with your CPO to talk through what's currently possible on the roadmap and what's impossible but, with the right adjustments lower in the product technology, might be available in the future.

A company's product roadmap doesn't describe what the company needs. It describes what the company needs limited by what the company believes Engineering is capable of. Your job as CTO is to simultaneously ship what the company needs and to improve what everyone believes is possible.

With a Shipped Potential chart you can get this done and optimize the output of Engineering. Without that chart you'll find yourself arguing over whether your company's airplane needs both wings now or if it's okay to build one and then come back and build the other "when we have more time."

Part Three

Technical Debt Financing

Your task is not to foresee the future, but to enable it.

— Antoine de Saint-Exupéry, in The Wisdom of the Sands

Technical Debt

It's time to plan for your company's future. In Part 1 of this book we went deep into what 'velocity' even means for Engineering, then in Part 2 we looked at the mental models that allow us to create a streamlined system. Now we'll address the actual drag and thrust that determine our forward motion.

Technical debt is one of the most powerful tools at your disposal. Your engineers may believe technical debt is bad and holds them back — it's certainly a cause of stress and a time sink — but I believe it only hurts us as an organization when we miss the larger picture.

The best definition of technical debt is "An obligation for future technical work." It's not "bad code" or messy data, though those might comprise it. With that definition, let's improve on conversations about technical debt which often make two crucial mistakes: They frame the debt as an ergonomic issue that primarily affects engineers (instead of a financial issue that affects the company) and they don't discuss it in the context of technical *investments,* interest rates, or ROI.

It's not just hygiene

Engineers are the ones who directly experience technical debt but this doesn't mean it's an issue affecting engineers; it affects the whole company. Your CFO might express worry about a bad contract the company got itself into but no one would say that's a problem primarily affecting the CFO. The CFO has merely identified the problem.

Technical debt is a contract that the company has signed. It's important to carefully pay these debts in order to free up Engineering's time for other work. The fact that it also reduces engineer annoyance is a bonus, but should not be the primary motivator.

This might seem like a meaningless distinction but it lets us be clear about where the debt comes from: Strategic decisions made by leaders at the company. Technical debt doesn't come from poor-quality engineering or the work of junior people, it's a contract the company signs to move fast now and go slow later. Debt appears when the company incentivizes (or allows) technical shortcuts for immediate value payouts. Leadership sets the standards for the work and implicitly communicates what kinds of timelines are most important (usually very near-term timelines).

So, since technical debts come from leadership decisions, we as leaders can change the debt strategy of the company as we wish.

Interest Rates

If we go 'fast now' and 'slow later' we should probably make sure the 'fast now' is worth it. Just like taking on a round of debt financing every technical debt has distinct terms and a payment structure. You could, right now, delete all the automated tests at your company. You would get a small but real increase in shipping speed for a few minutes. And then all hell breaks loose.

That's like taking a 12-hour loan from the mob.

You could also, right now, approve a new backend language to the list of languages your company supports. Depending on the language and the way it's used this might really help your teams. You'll have to support it, but perhaps it's worth it.

This is more like taking out a traditional bank loan.

As we think more clearly about principal and interest rates we're able to compare debts against each other. This is the first step to deciding which ones to pay down and which to ignore.

We do not want to eliminate all of our technical debt. Maybe someday, when the company issues stock dividends instead of reinvesting in R&D we can take that as a sign that there's no more useful product expansion and we should pay down our existing debts. Until then, we accrue debts so we can move fast.

We want those debts to be at an extremely low price, both in interest payments and payoff events, so we can use the capital (our newly-unlocked free time) to make big investments. Just like in the world of finance, we can't make a profit just by taking on debt; we have to purchase something with the liquidity that debt gives us. We make a profit when the thing we purchase is worth more than the debt.

Technical Investments

A technical investment is something that accelerates future work or earns future revenue.

For example, when we make deployments faster we earn a return every time we deploy, forever. When we migrate to a better-supported software framework then we have fewer edge cases that slow us down. When we reduce domain complexity or standardize the libraries we use then we add an accelerant to all future development.

These are all technical investments that reduce the cost of doing our work, and they're the kinds of classic tech debt initiatives you might see championed by senior engineers on your teams who are frustrated and know that it's possible to move faster.

Not all investments involve paying down the cost of doing work: Feature development is also an investment – the most obvious one. We create new value out of nothing and then we sell it and earn revenue (directly or indirectly, depending on your business model) forever.

If both cost reduction initiatives (like standardizing libraries) and feature development are investments, then we can prioritize them against each other. We can unify our mental framework of technical debt and technical investment into a more traditional financing portfolio. We'll look at that framework in detail in Calculating technical debt but first let's review how teams tend to address technical debt and why those attempts usually fail.

Common Pitfalls of Addressing Technical Debt

It's rare to see a sophisticated technical debt portfolio. Even companies that have a rigorous product research and feature prioritization culture might get a little hand-wavy when talking about debts.

More likely, the discussion of technical debt is limited to listing very annoying, bothersome things that engineers wish they had time to fix, with some of those hardest, most frustrating problems elevated to a kind of bogeyman that even individuals far outside engineering might hear about regularly.

It's easy to find agreement that technical debt is bad and less of it would be really good. When we try to prioritize specific debt is when the agreement is hard to come by — particularly if valuable product features need to be delayed because of it. Earlier in my career I presented to an executive team about pressing technical debt problems and had them all — from Marketing to Legal to Operations — tell me that they hear about the debt all the time and they sympathize, but that it wasn't clear how to eliminate the problem.

As I stood in the conference room with my slides up I realized I didn't have an answer for them. I, like most engineering leaders, knew the debts were a problem but I didn't have a plan that went beyond "the engineers agree we should fix this."

Actually trying to address debt tends to start off well-intentioned and then get stuck almost immediately in one of 3 specific non-solutions.

Failmode: 20% time allocated to debt

A popular one one is to attempt to carve out some time to address debt. Either declaring that one week a month is 'debt week' or 20% of time should, roughly, be allocated to debt. Some teams mark individual tickets as 'debt' and try to pull in a set number of debt points to each sprint.

This is a recipe for both low morale and ever-higher debt. Allocating 20% time to tech debt produces a kind of digital stagflation. Partly because there's no concept of the relative cost of different debts so there's no way to determine the best debt to prioritize. But also because there's no relationship between feature development and debt payments so the proportion of time allotted is entirely arbitrary. It's usually about 20% because that's as much as we can stomach losing from feature development, which feels like the more important work – even if the feature development is producing debt at a higher rate than we're paying it down.

Feature development is more important than debt payments along a very short timeline. But on a longer timeline, prioritizing mostly feature development will suffocate *future* feature development. As long as the team that's trying this 80/20 mix can only look one or two sprints — or even just one or two quarters — into the future then the debts will never actually be a priority. And since the debt is competing for the same work time as value-creating features, the only way the debts get paid is by grumpy teammates performing heroics; a surefire way to ruin a team.

Even if they ever do get the debt under control they go right back to making more debt, ignoring that they may have cleared the way for some high-return investments that are going unstaffed, particularly any kind of tooling that might prevent the high-interest debt from returning.

Failmode: A Technical Debt Team

Worse, a team might be formed with a mandate to pay down debt. This was one of the teams I led at Square back when we had a Rails monolith that was on fire. My teammates and I dove into the inferno and tried to tame the most central and somehow most neglected part of the company's product.

Unsurprisingly, it was really hard. But the worst part wasn't the work, it was watching other teams ship features quickly on top of the mess we were cleaning and feeling like we were somehow losing ground despite our efforts.

As a colleague of mine once put it, this "tech debt team" pattern is like sending one team to the office basement and another on a free cruise. The incentive and context mismatch is a recipe for relational conflict between otherwise friendly colleagues.

Failmode: Locally-visible debt reductions

This last pattern is the most intuitive one: Just letting each engineer and team fix the debt nearest them as they see fit.

The benefit of this is that the person paying down the debt probably has the most local context and is likely to move quickly. But there's no telling whether this is actually important debt. Is this engineer bailing out a sinking ship or are they rearranging chairs on deck while the whole vessel goes down? We return to the same prioritization problem: How do we figure out which debts to pay down with our finite time?

Leadership needs to provide direction on which debts to service interest on, which to pay off the principal, and which to ignore. And executive leadership needs to provide that direction over an articulable time horizon.

The bulk of this work is accomplished merely by facilitating a healthy, public, and ongoing discussion among your most senior engineers about the various debts across the company. The truth is distributed unevenly through the minds of your engineers and no individual has the full picture.

This might look like a periodic Technical Business Review meeting where senior engineers update a living document to reflect what they see in the system and spin off any important conversations. It might also be just a leadership culture of giving senior ICs the encouragement and time to explore problems and report on their findings. Either way, most of the debt-tracking work is accomplished by letting the people who can perceive the problems surface them.

That will give you knowledge about your debt portfolio and investment opportunities. But it's up to you to use that info to make company-level technical debt financing decisions.

Technical Debt Interest Rates

The debt financing metaphor applies much more broadly that we might expect. Not only are there debts and investments, but each debt has a specific interest rate and each investment has an expected return. This is what allows us to prioritize which debts to pay down.

Consider the situation where you owe money on student loans and also to a payday lender. The student loans are at 5% interest and the payday loans are at 100%. You get a surprise $1K as a gift, where do you pay the balance? Maybe split it evenly, $500 on each?

Under this scenario every dollar you get should go to the payday loan until you have it fully paid off. In fact, if you could somehow take out *more* debt at the 5% rate and use that cash to pay down the higher interest loan that would be wise.

The first step in considering which debt to pay down is just to figure out the rough interest rate. To be clear, this isn't an exact science. We're not going to get decimal-level accuracy here as we try to calculate our debt portfolio. But, just like using Big-O notation for algorithms, we can be directionally correct.

Losing Work to Interest Rates

In a high-interest loan very little of the amount we pay actually lowers the principal. The first payment services the interest and the next payment does the same.

This is one of the easiest ways to identify high interest technical debt: What is something that forces your team to do work and then forces them to do the same work again in the future? How much energy gets sucked up by this debt payment? And how long would it take to pay off the principal such that no payments were ever made again?

If you were to make a list of toil like that you'd have a strong start on a debt portfolio.

The sneakier debts are ones that require no ongoing interest payment but the principal is increasing continuously. This might be from forking a software framework that's now out of date, forcing your team to undo all of their work later in order to upgrade. Or building a software suite inside a monolith without building in boundaries that isolate major components of the products. In each of these cases you might have a surprise in the future where all work for a few months (or years!) is devoted fully to paying down a massive old debt.

Many of the debts in your system likely carry extremely low interest rates, despite how much engineers might complain about them. Maybe you've got a single page on your website on an old JS framework but it works and nobody plans to update it. It's not in the New® Hotness™ and it's an eyesore but if it's not an obstacle to your team as they work then it carries a 0% interest rate. It would take the same labor to fix it in 5 years as right now and it isn't a security attack vector. Which means you should absolutely ignore this debt.

Most debts aren't obviously 0% and they're not obviously giant scary ones – they're somewhere in the middle. My favorite way to calculate these is by thinking about your company's dimensions of scale and how each one might make the debt worse. For example, which debts grow along the axis of user growth? Which ones grow with engineering headcount growth? With feature count? With data size?

There are many things you can do to organize your debt portfolio but first you need to make it. You need an accurate accounting of the technical debts that matter to your company.

Calculating Technical Debt

The Five Properties of a Debt

There are roughly five things you need to know about each debt obligation in your system:

- **Principal** – What would it take to fully pay it off?

- **Interest** – How much energy is lost just putting up with it?

- **Increase In Principal** – How much bigger will the payoff be in the future?

- **Increase In Interest** – How much more energy will we lose in the future?

- **Payoff Events** – Are there any inflection points in the future that will necessitate a sudden payment in full?

You may notice I described the debts here in terms of creative energy, not just time. This is deliberate because technical debt costs us **cognitive drag**, not just time. Creative engineering work does not happen at a constant speed. An engineer might say only 10% of their time is spent on a repetitive task but if you dig deeper it's appears to be sapping a majority of their emotional and intellectual initiative. Toil like that can push the team to get distracted or snack on low-value work to avoid the more direct, frustrating work. Asking the team how much of their creative energy they *feel* like they're losing is a better measure of how much enthusiasm and insight is lost. Which is what we actually pay our engineers for.

To make these concepts clearer lets explore some hypothetical examples.

The Ballooning Postgres Database

One of your teams operates a service where all of the data is in a single growing Postgres database. The team notices that queries are getting slower as the storage size grows but this isn't a system where performance is critical. Your current setup supports database volumes up to 20TB of storage and you'll reach that maximum in a couple years if the current growth curve continues.

The team plans to shard the data and migrate to multiple smaller Postgres instances when they need to. They estimate it'll take the whole team about 3 months to do that – and it'll take longer if the data is larger.

The **principal** of this debt is what it takes to eliminate it: Three months of migrating to a sharded design.

The **interest** of this debt is how much drag it introduces to your engineers: None at all! Having this simple database setup has allowed them to move quickly on developing features.

The **Increase in Principal** is how much longer that sharding migration will take as the data grows. Let's say that your team believes that the work here is mostly code changes but the actual migration of data might take weeks of an engineer shepherding it if the data is close to the 20TB limit. Since most of the work is the code changes we'll say the increase in principal is low.

The **Increase in Interest** is zero because the the interest is and will stay at zero.

There's a **Payoff Event** where the full principal comes due: 2 years from now. Let's be conservative and say we need to have it in 18 months.

Principal	Interest	Increase in Principal	Increase in Interest	Payoff Event(s)
3 months * 1 team	zero	low	zero	18 months

The Asynchronous App on MongoDB

Another one of your teams operates a service that uses asynchronous code with callbacks on top of a sharded MongoDB cluster. The team complains about the difficulty of testing the asynchronous code and has had to invent and maintain a testing library to make this pattern accessible to new hires. The data they store is quite relational so they're frustrated with the document-oriented storage model. Much of their work is creating and fixing secondary indexes. They say, anecdotally, that 75% of their energy is spent just fighting the system.

The database sharding method has plenty of room to grow and the database instances themselves are reliable. But the team dreams of rewriting everything, even though they say it would take a full year.

To eliminate the **Principal** of this debt requires a full rewrite of the app and a migration to a relational database.

The **Interest** is the high percentage of the team's creative energy wasted by this debt.

The **Increase in Principal** is how much harder it'll be to fix this if we wait. If the solution is a rewrite that means this is growing in lock-step with the app's complexity.

The **Increase in Interest** is high because as this app grows there's yet more painful complexity to wade through.

However, there's no **Payoff Event** on the horizon. The team's output will decelerate forever but the system will technically keep working correctly.

Principal	Interest	Increase in Principal	Increase in Interest	Payoff Event(s)
1 year * 1 team	75%	high	high	none

Comparing two debts

Say one of your Directors of Engineering supports the managers of both these teams. The director comes to you and says both teams are asking for time to pay down their debts. What insight do you offer?

If we do nothing then in two years the Postgres database falls over and that's a total outage. So we have to do something about that. Whereas the MongoDB team just gets sadder and slower but their system keeps working. So should we ever prioritize letting them do their rewrite?

That depends on what the purpose of that second system is. What benefit does the company get from this painful asynchronous app? Is the low morale of the team offset by a good feeling that at least their work is extremely important? Or is it a trivial set of features that could be deleted?

Assuming both systems are roughly equal in their importance to the company, I would advise this director to stem the bleeding on the async MongoDB system by curtailing new development in it.

We can cap the growing principal and interest if we don't add new features into that system. To do this, we'd need to staff a one-time migration to design a replacement system that new features can go into and write a full proposal on (eventually) migrating all existing features to the new system. (It's important to write that detailed proposal in order to test whether it's possible to one day fully decommission the existing system.)

Considering the timelines involved, I'd advise this director to give the next year of time to the MongoDB team to build and start to use a new approach, then spend the second year focusing on the sharded Postgresql migration. If the company hasn't totally transformed after that, the third year the MongoDB team can finish decommissioning the original app.

This is an imperfect science, but as long as high interest rates are tracked and addressed the team should be able to recoup their creative energy.

A Notation for Scale

Note how in the 'Increase in Principal' and 'Increase in Interest' columns for the second example I put the word 'high'. That's not very helpful. How do we compare one 'high' against another? What if most of our debts have 'high' interest?

Let's look for a better way to describe how debts can get worse over time.

If you've ever interviewed at a company that hasn't updated their hiring philosophy since the 1990's you might have encountered Big O notation. It's a way of describing the worst-case scenario of the performance of some logic as the logic is applied to data. Let's steal just a piece of this concept to describe the way that debt can get worse over time. Instead of using a single dimension 'n' as in '$O(n)$' versus '$O(n^2)$' we'll look at all of the different dimensions along which a digital system can grow and therefore debts can get more expensive.

Dimensions of scale

This book is focused on engineering teams that support online software systems because that's been my whole career. These systems grow along many axes over time: You'll get more traffic, you'll store more data, and the graph of your data relationships will become more complex. Alongside all that, you'll see more engineers working on it, more customers using it, and an ever-larger range of dates represented in the production dataset.

You may have other dimensions of scale, depending on your business model.

User Traffic

This is a classic measure of scale. Creating a version of something that ten people can use at once is phenomenally easier than one that a million can use at once, independent of the dataset behind it.

Data Storage

Another classic measure of scale, there are some cliffs to watch out for here (like running out of storage for any non-distributed database) but even incremental growth here will cause your queries to respond slower and your hardware expenses to go up.

Feature complexity

I love keeping track of the number of supported features. Partly because it gives us the Shipped Potential chart but also it's helpful to know roughly how many different user experiences the system supports. This kind of inventory makes it possible to answer the question "If we build X how many existing features might need to be adjusted?" In practice, a question like that can surface a rough coefficient for multiplying the back-of-the-napkin time estimate that an engineer might give.

It's not rare for each feature to take longer to develop than the previous one. So, for the purposes of modeling technical debt, it's helpful to use feature complexity as one dimension of scale along which a debt might get more costly.

Data Modeling Complexity

Most of the companies I've worked with could pay more attention to this dimension. This is a measure not just of the number of databases, tables, and columns that exist in the schemas at your company, it's a measure of overall graph complexity.

If you were to generate a diagram of your production schemas and data relationships it might be super ugly. Instead of a neatly organized tree structure you'll probably find a few datasets that virtually everything references. These datasets would also be the most painful ones to work with, they'd represent the central concepts of your flagship product, and they'd probably have way too many fields in each dataset.

If you've never calculated the graph complexity of your production schema before, there are plenty of different measurements you can make. To start, I recommend keeping it very simple and just counting two metrics: 1) The p90 and p99 number of columns in all tables, and 2) The number of relationships between tables divided by the total number of tables.

Number of Employees at the company

Is there manual work for your technical team every time one more person joins the company? Or are the administrative interfaces that employees use to manage and operate the product starting to creak at the seams?

This tends to only matter for administrative tools, but it doesn't look great when an engineering leader is totally surprised by a scaling cliff for tools their colleagues use. And if you find that more than half of your employee headcount is customer support you wouldn't be the first leader to be startled by that; many startups that fail to prioritize internal-facing debts have to hire budget-destroying Customer Support teams.

Number of Engineers

You'll likely experience slowdowns in development as your engineering headcount scales but it's not necessarily related to technical problems. Here you'll find process debt, cultural debt, and communications debt in addition to some technical debt.

As engineering headcount scales you may find brittleness in your build and deployment tools, especially if any special knowledge might be required to perform a deploy. At some point you may also notice the scarcity of good code reviewers because as headcount scales up the percentage of the system that any individual engineer understands goes down.

The debts here are almost entirely non-technical. I recommend separating those debts from the strictly technical ones for the purposes of your technical debt calculations. The process and cultural and communications changes don't compete for your time from the same work queue as the technical ones. Nobody's going to ask you to choose between shipping a feature or overhauling your internal communications. The expectation is that those two pieces of work can happen in parallel.

The one place I recommend looking very carefully with regard to engineering headcount scaling is Not Invented Here syndrome. Where in the product is there an invention that doesn't absolutely have to be there? Notice which debts don't work as well with lots of new hires. Too often a technical system relies on an unnecessary invention understood by a select few – often the same few people who're needed to do other critical work. This is especially dangerous if the invention was written by a technical cofounder who's now in an executive role of some kind because their invention has likely been (unconsciously) shielded from scrutiny.

Number of Users

This is a big one. Regardless of data size or throughput in bytes, features that worked for a hundred users rarely work for ten thousand. Administrative interfaces will get slow, synchronous workflows will need to be made asynchronous, perhaps the primitive search system for exploring user data will need to be fully replaced, etc. And any piece of code or UI that operates across multiple users (typically found in analytics jobs and admin interfaces) will strain as user count goes up.

Perhaps more sneakily, the number of edge cases that your team will see in the data relationships scales roughly in line with user growth. Given enough users, you'll see every possible permutation of user data, each of which will need to be encoded in the test suite using ever more complex data in the tests.

Each passing day

There are some systems that record a snapshot in time or perform date calculations over a range of the full dataset. Even when nobody's using these systems this dimension of scale can get worse merely from the ticking of the clock.

Using Dimensions of Scale

So instead of saying that an interest rate is 'high', we can say that it gets worse along specific dimensions.

In the case of The Asynchronous App on MongoDB the principal of the debt got worse with feature growth and with relational data complexity. And the interest got worse with each new engineering hire at the company. Even if the engineers on the team felt like a steady 75% of their time was wasted slogging through, for each new engineer at the company there's a greater bifurcation of architectural approaches and increased desire for people to leave this team and work on something better.

We can describe this debt obligation a little better now.

Principal	Interest	Increase in Principal	Increase in Interest	Payoff Event(s)
1 year * 1 team	75%	features * data complexity	engineer count	None

And in the case of The Ballooning Postgres Database we saw that the principal increases as data size increases, so we can be more specific about that.

Principal	Interest	Increase in Principal	Increase in Interest	Payoff Event(s)
3 months * 1 team	zero	data size	zero	18 months

With that, let's make a technical debt portfolio for your company. We'll assume you have to deal with both The Ballooning Postgres Database and The Asynchronous App on MongoDB. On top of that, let's contrive a few other realistic debts for your teams.

Yes, these are all situations I've lived through and, yes, forcing you to hear about them is absolutely therapeutic for me.

The Slow Admin Dashboard

You've got an internal administrative dashboard app that lets employees navigate user data. It's existed for years and the pages are getting slow. It has its own permissions to read and write to the databases that sit underneath the product applications and it queries them directly. It only performs 'SELECT' queries but they are very inefficient – on average a page loads in about 30 seconds.

You're worried that one day too many of your colleagues will try loading the dashboard's index page at the same moment. The page is so data-rich that even a few dozen simultaneous page loads can lock up one of the production databases and cause an outage. Your engineers want to replace it with a new client side app that reuses existing HTTP endpoints in the product instead of direct access to their databases.

There are actually two debts here. There are two things that need to be done, so there are two obligations for future work. One is that you'll need to start using read replicas for this dashboard app. Simply loading a page should never cause a production outage and the best cache for a database is a database's read replica. This debt has a **Payoff Event** that's imminent. The other debt is that you'll need to move away from direct shared database access as an architectural pattern.

Let's make an entry in our debt portfolio for each of those.

It'll take about a day to move the queries from the primary database to a read replica and payoff the **Principal**. There isn't any repeated maintenance labor caused by this app (other than perhaps a looming dread) so the **Interest** is zero. That won't change so the **Increase in Interest** is also zero. And the cost to switch to a read replica is the same both now and later so the **Increase in Principal** is zero.

Fixing the architecture of this whole app is a far, far harder job. The engineers say they can do it in 6 months but maybe you've seen this before and know that the **Principal** here will take a full team 2 years at a minimum. Since fixing the architecture is effectively a rewrite you see an **Increase in Principal** with each new feature that'll need to be ported from the old way to the new way.

The slowness of the app causes all employees to pay an **Interest** payment of about 30 seconds every time they use a page. And it's getting slower so you see an **Increase in Interest** as database size grows, as the user count increases, and as more employees use the pages.

To make it all worse, the load balancers for this app have a maximum 60 second timeout so as soon as these slow pages can't respond in that window the app suddenly stops working completely. Looking at some charts you estimate that **Payoff Event** will happen in 9 months.

	Principal	Interest	Increase in Principal	Increase in Interest	Payoff Event(s)
Admin needs a read replica	1 day	zero	zero	zero	any day now
Admin needs HTTP APIs	2 years * 1 team	Waiting 30 seconds to see any page	Features	employees * data size * user count	9 months

Both of these are critical. The second one, at least, won't fail immediately. So the right course of action is to fully pay off the first one as soon as there's a good moment in the team's cadence and then work on a thoughtful strategy for addressing the second – especially considering the payoff event is sooner than the principal payoff period.

Engineers Cloning Production

An early employee created a script that dumps a production database directly to standard out and pipes it into their laptop's local database. It's extremely popular because any engineer — particularly junior ones — can download production data to their laptop and run the app in development mode to see how their code changes perform in production.

You hope to raise a round of funding soon and you know that that'll require an audit of your security and compliance. This will be flagged as a major compliance violation because user data should never leave the production environment. And should definitely not be sitting around on an employee's laptop – certainly not the *entire* user dataset.

You also know that this poses a handful of other major problems for your organization in both engineering culture and technical sophistication. Here, I'm going to talk about the nuance of this debt but in case your company is leaning toward this pattern let me urge you in the strongest possible terms not to do it. Your teams should have production-like data in their unit test fixtures. Until that's true the product development quality will be low and there'll be a temptation to download data from production.

Calculating this debt is interesting. The debt isn't that bad right now, though there are some **Interest** payments in the form of lower security, script maintenance, and the lack of comprehensive test fixture or test factory data. But there are several **Payoff Events** and when we suddenly need to move away from this approach it will take an unknown amount of work to pay the **Principal** and get our system in shape to be developed through better patterns.

There's an **Increase in Principal** as more code is written with insufficient unit tests or poor service boundaries, and each new engineer who joins and uses this pattern puts us further behind.

There's an **Increase in Interest** with each new engineering hire as people start asking for slight improvements in the script or for better production data scrubbing and development time gets allocated there.

	Principal	Interest	Increase in Principal	Increase in Interest	Payoff Event(s)
Cloning production data to laptops	unknown	5%	features * Engineers	features * engineers	Any audit, Hiring a staff engineer

Messy core product modeling

For the purposes of this example, let's say your flagship product involves file storage. This is such a key concept at your company that the word 'file' appears in most conversations and is in the center of most technical whiteboard drawings. There's a class called 'File' in the main app and it's out of control: thousands of lines, relationships to most other classes in the codebase, and with what appears to be two poorly-implemented finite state machines within the class itself.

Paying down the **Principal** here is just a ton of work. Hundreds or even thousands of careful refactorings, moving logic out of this class into something better encapsulated and modeling File-related concepts in their own classes. Your team pays **Interest** every time they touch this file or an adjacent one and, anecdotally, it takes a week to do in 'File' what could be done in an hour elsewhere in the system. Every time new features are added there's an **Increase in Principal** and an **Increase in Interest**.

How would you calculate the actual interest rate for File? Taking a week to do an hour's worth of work sounds extreme – that's a 40x slowdown whenever work gets close to this code, or 97.5% of energy going to servicing interest payments. But not all the work happens here, most of your team is working elsewhere in the system so it's not like all of engineering is slowed down 40x. If it were, this would be the most critical piece of debt to pay off before anything else.

This situation is common and also a good reason to have a finished debt portfolio. Because it's impossible to know whether to pay this down without knowing how many File-related features are on the roadmap and what other debts are also in the way of upcoming work.

To calculate this debt let's just get a rough sense for how much time engineers will spend in this are of the code and make a guess at how much this debt might frustrate them. Maybe your upcoming features will be half of the speed they could be (a 50% interest rate) because a few of them require working in this debt. Rather than agonize over the exact baseline make an educated guess. If you're wrong it'll come up in discussion with your senior ICs.

	Principal	Interest	Increase in Principal	Increase in Interest	Payoff Event(s)
Messy core product modeling	years	~10-50%	features	features	no

A Full Technical Debt Portfolio

Let's put all these contrived but believable examples together and see if we can compare them against each other.

	Principal	Interest	Increase in Principal	Increase in Interest	Payoff Event(s)
Ballooning Postgres Database	3 months * 1 team	zero	time	zero	18 months
Asynchronous app on MongoDB	1 year * 1 team	75%	features * data complexity	engineer count	None
Admin uses primary database	1 day	zero	zero	zero	any day now
Admin reads from database directly	2 years * 1 team	Waiting 30 seconds to see any page	Features	employees * data size * user count	9 months
Engineers Cloning Production	unknown	5%	features * engineers	features * engineers	Any audit, Hiring a staff engineer
Messy core product modeling	years	~10-50%	features	features	None

It won't be clear and obvious which of these to work on and it what order unless we know what we're trying to accomplish next. Well, perhaps moving the admin app to a database replica is a shoe-in for prioritization because of the low cost and immediate payoff. But prioritizing the rest depends on what features the company wants next.

Will your teams be working exclusively in the 'File' class? If so, better make a plan to either clean that up or somehow mitigate the worst parts of it.

Does the company critically need to improve customer NPS and lower customer service headcount growth? Perhaps a super-responsive and more powerful admin dashboard is a must-have.

Are you planning to double engineering headcount in the next year? If so, both the asynchronous MongoDB situation and the production cloning pattern might need to be remedied immediately.

The point of a portfolio isn't to make a TODO of fixes but to expand conversations to include the entire company's technical context.

Taking the Executive View

Most of the time, leadership isn't stepping out front and saying "This is the way, follow me!" but it does frequently involve giving a map and a route to your people. Even if you don't know how to draw the map, as the leader you need to source it and provide it.

Remember we're trying to minimize drag here. Cognitive drag, the energy-sapping confusion and conceptual graph complexity that an engineer feels when trying to be successful in a complex environment.

As an executive leading a modern software team your people face huge complexity. The territory in which your engineering teams work is literally interconnected concepts written down. That's what software is. Leadership, here, needs to be some kind of guidance through that maze of concepts that allows your engineers to put only the necessary concepts in their head to navigate from where they are to where they need to go.

A technical debt portfolio is a map of conceptual territory along the dimension of time. It marks where there are impassible obstacles, where there are arduous hills to climb, and where there are smooth paved roads. While an architectural diagram can describe where you stand in that territory right now, a debt portfolio gives you insight into the safe roads for the journey ahead.

Part Four

Empowered Teams

One will weave the canvas; another will fell a tree by the light of his ax. Yet another will forge nails, and there will be others who observe the stars to learn how to navigate. And yet all will be as one. Building a boat isn't about weaving canvas, forging nails, or reading the sky. It's about giving a shared taste for the sea, by the light of which you will see nothing contradictory but rather a community of love.

— Antoine de Saint-Exupéry, in Citadelle (1948)

Yearning for the open sea

We've built our streamlined airplane and we can plan a route. Time to fly it.

This book is all about reducing the things that drag on our engineers while maximizing their thrust. To that end we work at a higher, more inspiring altitude. If we instruct our people to build a ship they'll eventually churn out something that floats but if we help our engineers develop that "shared taste for the sea" then it's more likely the work will go swiftly, happily, and correctly as they produce something elegant and right.

In our case, that shared taste is for positively impacting another human through our work. Making technology that improves someone's day, somewhere. When we make visible the connection between one engineer's labor and some other person's smile we create teams that work quickly, driven by their shared taste for the sea, as it were.

In this part of the book we'll outline four policies that drive toward that goal. Four policies that remove drag and provide thrust so a team can understand it's **impact** on users and giving them the **focus** and **process** and **information** to get there.

Focused Experience Teams

 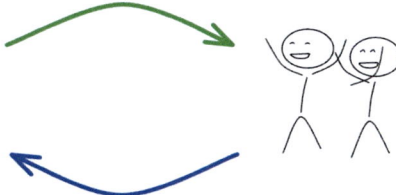

The four ways to charter a team

We can discover whether our teams are aiming directly at the right problems by clarifying their charter[22] (what we commission them to do) and then evaluating how their planned work lines up with the stated purpose of the team.

There are basically four ways a team can be chartered, with varying levels of usefulness.

1: Time-Bound Project Teams

The worst, most regressive way is to create a team that performs a short-lived project. These teams are often doing extremely important work and may comprise the most experienced folks in the organization – precisely the people who need to do this work. But it's not a team. It's just a group of people working on a specific problem. Once the problem is gone, the team makes no sense. They have no opportunity to go through Forming, Storming, Norming and Performing – they just do the work and then disband, wasting any investment a leader might have made in team camaraderie or group dynamics.

For a two-week project a temporary team can be fun and useful. Longer than that requires a structure that can handle vacations, sick days, mentoring, and all the attributes that make real teams powerful. If this team is getting disbanded at the end of the project then we'd be better off breaking the work into pieces and assigning it to existing, long-lived teams.

2: Technology-Centered Teams

Only slightly better is to have a team do work that fits a specific area of technology. Maybe they work on a legacy framework or a particularly troublesome datastore or just a grab bag of otherwise-abandoned systems. The work goes to them because someone identified a problem and routed it to them based on where that problem came from, not because of which user had

---

[22] Some companies let teams write their own charter. I recommend against this — it's better for the leaders who support the team to write it as a record of minimum expectations.

the problem and which engineering team serves that user.

These are also rarely teams in any useful sense. They tend to attract technological hobbyists and people who are (possibly unknowingly) committed to the existence of the technology they work on. If you were to find a way to delete this whole part of your system and replace it with a free 3rd-party library they might feel threatened rather than celebrate the efficiency gain.

These teams are prime candidates for re-chartering toward a user-focused purpose.

3: Product-Shipping Teams

Better than that would be to charter the team to launch a specific feature or product. This approaches our ideal because the team is focused directly at real user value.

But even here there comes an inflection point where, when the thing is launched, the team switches modes. The team has spent its entire existence trying to create something novel and now, the moment it's launched, they're a reactive team that addresses user confusion, bugs, incomplete features, missing features, and all the incident response that comes with it.

This is a two-phase team that functions kinda like a stem cell: It exists first in a really high-potential form and then turns into a specific, permanent thing. And it stays there forever. With this model each time we launch a product we either lose the team to permanent maintenance or we ask the team to abandon the product and move to the next one.

We can't afford to add a new team for every big launch.

4: Focused Experience Teams

Which brings us to an approach that keeps teams focused on doing real work for real users, wherever it appears in the product suite: Facilitating a specific experience for a specific audience.

The most engaged, valuable, and joyful teams I've seen have been teams that know precisely who their users are and understand how their work solves the needs of real humans somewhere. The audience might be just prospective new users, churned users who we want to come back, internal admins, engineers on other teams trying to run tests, 3rd party development teams, etc.

And the most effective and focused of those teams are able to articulate specifically what experience they facilitate for those users.

A *Focused Experience Team* is a team that's chartered to serve an articulable user group of some kind with some valuable, technology-leveraged experience. Every team in Engineering can and should be a Focused Experience Team.

If you're familiar with the Jobs to be Done framework this has a similar positive effect of forcing the work to happen through the lens of some human's felt need. If the need is small or the user in question doesn't seem to ask for the experience then we have to question the existence of the team.

These four team charter models we just stepped through have varying levels of emotional connection between the engineers on the team and the end user. This connection is the key – it's the yearning for the vast sea that makes ship-building itself a satisfying labor.

The impact of this is easiest to see in the most junior engineering hires. Early-career engineers are not yet desensitized to the power we have to craft someone else's experience. It's not rare among junior engineers to hear stories of them sharing feature launches with their parents, partners, and children — you can sense their pride in the impact.

Focused Experience Teams are a tool that leadership uses to augment that pride. To encourage the emotional connection between engineer and end-user that inspires engineers to push obstacles out of their path.

Common Focused Experience Teams:

You may have already worked with teams that have a Focused Experience Team charter. A "Developer Experience" team is a common pattern of a team chartered to support a specific (internal) user experience. There are plenty of others as well, sometimes hiding in plain sight.

Let's take a minute to see how many common teams either already are Focused Experience Teams or can be very easily rechartered to point more directly at a user experience.

- *Growth Engineering* – serves prospective users aren't signed up

- *New User Engineering* – serves users who have signed up but not performed all actions that we believe correlate with retention

- *Data Engineering* – serves internal decision makers and Data Science by producing correct, timely data products

- *Data Science* – serves internal decision makers by producing guidance

- *Internal Product Engineering* – serves all employees with an admin interface that connects to the whole product suite

- *Finance Engineering* – serves the Finance team at either large companies or at startups where financial reconciliation is related to the business

- *[Server] Infrastructure Engineering* – serves engineers with internal APIs that abstract away data processing and sometimes higher-level product primitives

- *Web Infrastructure Engineering* – serves engineers with UI frameworks and libraries that reduce development time

- *Security Engineering* – serves engineers with secure libraries & systems that reduce development friction

- *Deployment Team* – serves engineers with deployable targets and a deployment workflow

Leading Focused Experience Teams

Budgeting

Focused Experience Teams make it easier to set a team headcount budget. The team exists to serve a defined customer, that customer has some value to the company, and you so you can set a cost target for serving that customer in lockstep with creating an engineering headcount.

To set any investment budget you must know two things:

1. How much value can be realized from an investment

2. The curve of how much cost yields how much value

That lets you select the point where you'd like to invest. Most investments are somewhat sigmoidal such that the values only change in the middle: A small investment doesn't get you much and a overly massive investment is the same as a moderate one.

During the the 2010's when interest rates were zero it was popular for tech companies to add headcount at the far end of this curve where half as many people might have yielded 80% as much payoff. During more realistic markets it's important to be careful about where we are in the curve.[23]

Let's look at some examples of using this budgeting approach with Focused Experience Teams.

---

[23] There's a calculus problem here where a single point has a derivative of zero and we want to invest right there.

Executive Engineering

You may have a 'Developer Experience' team that accelerates other engineers; their customer is all of Engineering. If you have, say, 100 engineers then making Developer Experience a team of 1 is likely too small and a team of 50 is much too large. You probably have a mental model of the curve from the image above and how it works within Engineering to accelerate engineers. There's some point on the curve where it feels more worthwhile to stop adding engineers to Developer Experience because you'd rather have them shipping product.

Now apply that same thinking to a Customer Support Engineering team whose customers are your colleagues answering support requests from strangers. There's a curve here, too, except the headcount is drawn from two different budgets: One for Customer Support personnel and one for engineers.

Step one of setting the engineering headcount budget is deciding how much value can be realized from this investment. The maximum value might be "laying off all of Customer Support because it's automated now." More likely it's limiting Customer Support headcount growth to, say, half of the current growth rate. Even that might be a huge financial win.

Step two is figuring out roughly what the investment curve looks like.[24] What could a single engineer accomplish, what could a large team of engineers accomplish, etc. Even fake numbers here are more useful than not thinking through the problem.

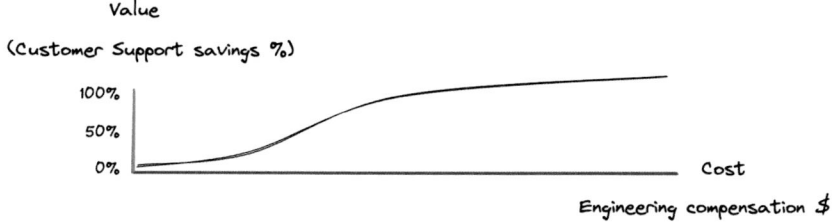

---

[24] I was first introduced to this pattern by Peter Seibel from his work on how one team can scale another: https://gigamonkeys.com/flowers/

Now you can model the savings along a timeline and see roughly how much you might save with different levels of investment.

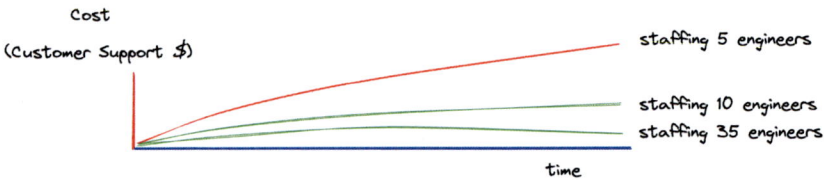

The right choice depends entirely on the financial realities of your company and the other investment opportunities available. For most companies this isn't even a choice, they just pay what the Customer Support burden happens to be and they may cut budgets elsewhere in a pinch. Focused Experience Teams allow us to model future ROI tradeoffs and budget in advance for each team.

Layering Focused Experience Teams

This budgeting is helpful when we're using Technical Coherence to design our org. The three layers of teams don't necessarily map to the reporting hierarchy in the org chart[25] so we can't use some of the more common patterns[26] to allocate engineers.

Let's look at the teams for an example company with about 150 engineers:

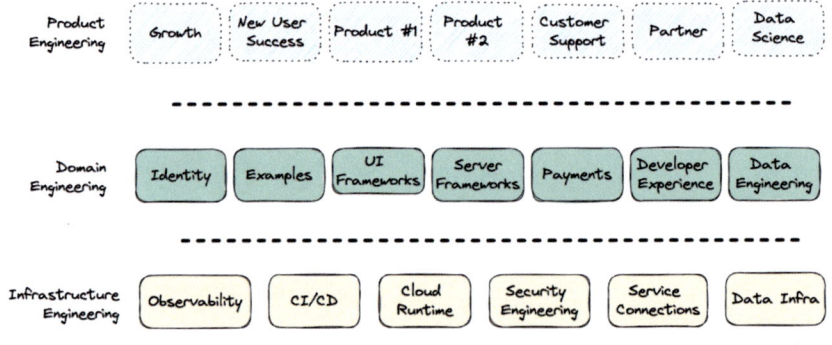

---

[25] It's often useful for a senior leader to lead both Product Engineering teams and the Domain Engineering teams that primarily support those Product Engineering teams.
[26] many of which are a thin veneer on empire building.

At the top we have Product Engineering teams serving specific UX domains like 'New User Success' and external partners. We can immediately recognize the customer of the Product Engineering teams. Each one is a Focused Experience Team for users performing a very specific role.

It's harder to see how the Domain Engineering teams are Focused Experience Teams because Domain Engineering teams, as a set, support the whole layer above them. Let's examine a couple of them closely to see how it's done.

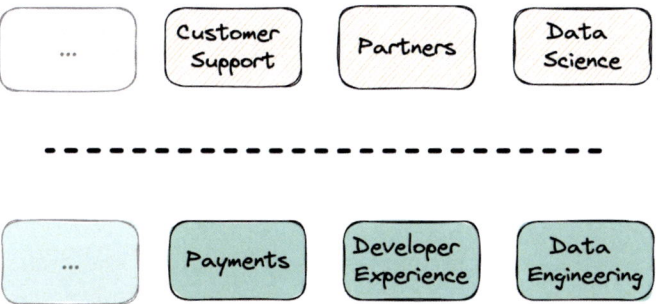

Each Domain Engineering team provides a specific experience. The most obvious is Data Engineering which provides data for Data Science but also provides APIs and tools for other engineering teams to produce usable data products.

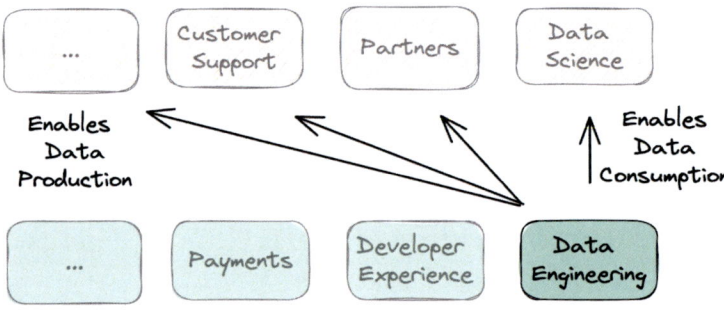

Data Engineering as a Focused Experience Team enables the *experience* of moving data around the company. Anyone who needs to do that is a customer of Data Engineering. A common Data Engineering failure mode is to lose track of these experiences and focus on the technical work. Completing big migrations on time or using the latest tech doesn't matter if nobody has access to the data they need.

The same pattern works for other Domain Engineering teams. Developer Experience could be a team that makes it easy to add and iterate on product features. Both of those examples are a little generic because they don't encapsulate novel competitive advantages in the product, so let's look at one that does.

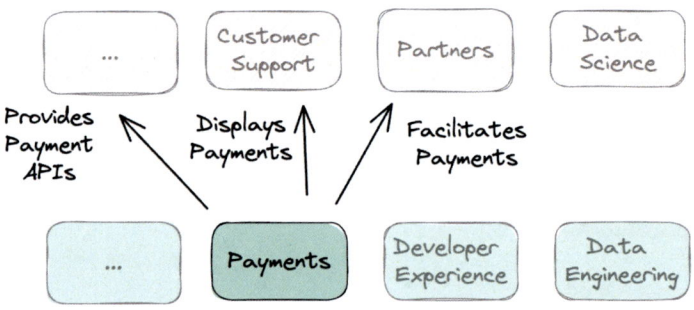

Let's imagine a company that offers some kind of checkout flow. Payment code is usually some of the most feared[27] — and therefore unmaintained — parts of a product suite due to its sensitivity. So we could ask Product Engineering teams to hack at it from time to time when they're adding a feature but it would become a big unsafe ball of debt. Or we could staff a Domain Engineering team to abstract away all the complexity and provide the necessary payment experiences. High-level abstractions that folks can use when adding payments to features, a third-party interface when necessary, a solid audit log and visualization for Customer Support, structured data for analytics, etc.

If payments show up in multiple UX domains then staffing this team at the Domain Engineering layer will reduce the drag on multiple feature-shipping teams. And staffing it as a Focused Experience team will keep the team thoughtful about why they exist and what jobs need to be done.

Focused Experience Teams at the Infrastructure layer work just as well and keep the Infrastructure roadmap focused on real value.

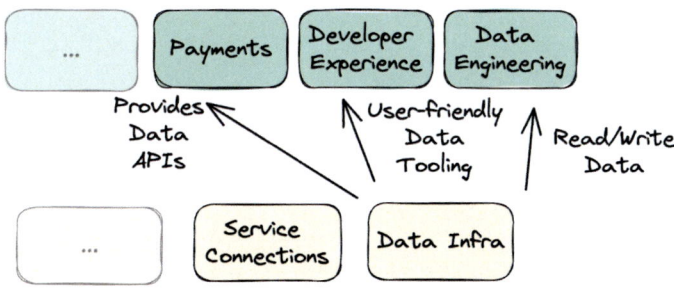

The Data Infrastructure team has customers with specific needs. When we make Data Infrastructure a Focused Experience Team we suddenly get clarity on what technologies *not* to use. Perhaps Data Infrastructure wants to design a massively scalable data system. Most startups — even to series D and beyond — have so little data that all of the important production records could fit on a

---

[27] This even happened at Square, a payments company. The first few versions of payment processing were the most confusing and least-touched code.

modern phone. So the right choice for Data Infrastructure is to build well-abstracted APIs that provide the right experiences and then implement those APIs with an appropriately-scaled too somewhere along a spectrum from SQLite to Kafka. The choice will be clear only if they keep their customer in mind and plan their work to deliver valuable experiences immediately.

Localized Decision-Making

When we pay so much attention to the experience a team provides we pay less attention to the implementation details of how they provide that experience. This can lead to some weird outcomes like creating redundant technologies or deviating from the engineering culture.

This means it's now even more important for you, as the leader, to ensure there's a coherent technology policy that allows for local decision making without a proliferation of different tools and approaches. Be sure to have a technical policy that enforces only one technology solution per problem type and give Infrastructure teams a mandate to provide one great development path for each pattern.

This is where you'll expectations for when, say, a Domain Engineering team should run their own infrastructure in order to provide the right experience versus forcing them to use what Infrastructure (who should be mindful of their customers' needs) has available.

Engineering, not Developing

Allow me a brief aside.

You may notice I'm consistently using the term 'engineer' in this book instead of 'developer'. The word 'develop' comes to us from Old French where it meant 'to unroll'. Developers unroll an idea from a ticket into source code. That's a small job fit for consultants or development agencies.

Engineering is the full ownership of a user's problem and its solution, from the design stage to completion with a guarantee of continued functionality in the future. That's a large job and if you give your teams the mandate to do it you'll get more out of them and get to watch your colleagues grow into their full potential.

This is the hidden reason why Focused Experience Team charters work so well. They invite the team members to a higher calling. Away from merely implementing tickets and toward providing a correct, fast, secure, technology product that solves real user needs. It's a way of trusting the team to deliver for the company and to do so using their full set of skills as empathic, intelligent humans who want to do good.

But, to keep with the theme of the book, it also results in better software shipped faster. It aligns the team with the user, reducing drag. It engages their autonomous drive for impact, increasing thrust.

Making this part of your culture

This might all sound good but how do you actually do this?

Rolling out Focused Experience Teams is both a cultural change and a policy change. The culture is harder, so let's address that first.

It's important to clearly articulate the cultural norms. I typically do this in a quarterly 1-hour presentation to all of Engineering where I review our values, summarize our progress this far, and communicate an updated roadmap. I'll generally spend ten minutes diving deep into a specific policy or cultural norm to ensure it's extremely clear.

You can start by declaring what you believe every software team should have. Focused Experience Teams are a team chartering model that imply certain norms. But you can also just say them plainly. For example, you can tell your org that you believe every software team must have:

1. A software-leveraged user experience they're responsible for enabling

2. Full control over their technical decisions, within the constraints your leadership sets for the whole org

3. The right and the time to work themselves out of a job

4. Access to all the information they need to accomplish their work optimally and get and give feedback

We defined these user experiences as part of Technical Coherence, the technical constraints should be set by teams in your Infrastructure Engineering layer, we addressed the planning structure with Technical Debt Financing, and we'll soon cover Communication Design.

Turning this into a Policy

To immediately roll out Focused Experience Teams as a policy, I recommend rewriting each team's charter (whether there's a written one already or not) to center the actual user of the team's work.

Offer the team something like the following template and have them fill in the blanks, then use this as the start of a document that outlines how the team works, where the team wiki is, etc.

(Use the upcoming Competency-Based Agile for setting a "How We Work" section of the charter)

Template for Focused Experience Team Charters

```
The {team name} team provides {identified user type} with
the technology experiences they need in order to {what
the business wants that user type to be able to do}.

We depend on {list teams at the company who produce data
or code that this team depends on} and we give them the
best visibility we can into our roadmap.

We're a dependency of {list teams at the company who
reference data or code that this team owns} and we
collaborate with them to ensure they're not blocked.

We evaluate our success by {either user NPS or something
the company appreciates more} which is visible at {insert
dashboard link}.
```

Competency-Based Agile

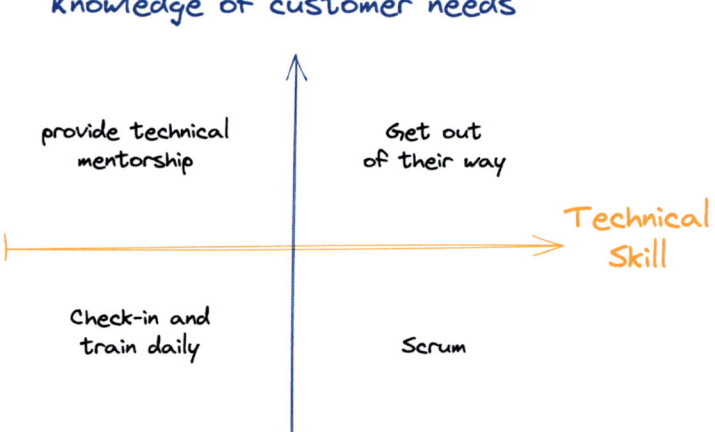

Creating Team Process That Works

In my leadership career I've failed at least two teams. Both teams tried their best but didn't reach their goals and it was largely my fault.

One was an Infrastructure team that followed a strict Agile workflow. Weekly sprint planning, weekly review, tickets, the whole lot. We talked about the work constantly but weeks would go by without much forward progress. We had no trouble figuring out precisely what we wanted to do; the team just couldn't make progress.

The other was a Product Engineering team with the exact opposite problem: We skipped a lot of the Agile ceremonies and instead focused on pair programming, technical mentorship, and design review. The team made some progress but never seemed to achieve anything. Week after week they were miserable, feeling like they were running in place.

Only later did I see the pattern: Each team had fallen into a development process that might have worked for somebody but sure didn't work for them. The Infrastructure team knew what to build but lacked the technical skill to execute. The other team was the reverse: They had plenty of technical support but didn't know what to do.

Levels of Knowledge

when the engineer is the customer — **Full**

↑

when the engineer can internalize the customer's needs — **Medium**

↑

when the engineer cannot access the customer — **Zero**

In 2001 the Agile Manifesto consolidated and popularized the best wisdom of software consulting. The document itself is full of good ideas and most of them are now best practices across the industry. Some of the ideas, like holding regular team retros and ensuring direct interaction between engineers and stakeholders, are valuable regardless of the particularities of the company. Others, especially the project management parts, are only useful in specific situations.

A client approaches software developers with a need and pays for a solution. If they don't like the delivered solution then either the client gets worthless software, the developers don't get paid, or the developers have to rewrite it. Any of those outcomes is a financial tragedy for one side or the other. Agile is a really thoughtful way to de-risk this process.

So if you don't know what to build, aren't paying for it to be built, won't have to own or operate it after it's built, and run the risk of (negatively) surprising a client then Agile is definitely right for you. None of that applies if you're, say, a security engineering team at a FinTech startup upgrading a rate-limiting architecture for internal services. If your problem isn't a lack of knowledge about what the user will buy then Agile provides less value to you.

The two teams I mentioned earlier had a development process that didn't fit them. The point of a development process is to remove risk and there are two major risks a team faces while making software: That a team won't know what to build and that it won't know how to build it.

Let's start with the 'what' because that's where Agile can be useful. We want to reduce the risk that we build the wrong thing. Something that, no matter how technically perfect it is, provides no value to anyone willing to pay for it. We mitigate this risk by gaining knowledge about the market, the end-user, and their needs so we can develop something that they'll definitely value.

The discussion here usually goes no farther than comparing 'Agile' to 'Waterfall' with a wide range for what those terms mean. A better discussion starts with identifying how much knowledge we already have and crafting a process to fill in the rest.

Knowledge as Proximity to Revenue

How much knowledge the engineering team has about its customer depends on who measures the value of the output. Who's the decider about whether a feature has been built correctly? Is it millions of strangers on the internet? One enterprise client? The CEO?

We evaluate our consumer products by whether consumers are using them (in a way that generates revenue) so for this business model the consumer is the decider. For other business models there's often an identifiable purchaser. And in highly regulated domains only a trained specialist knows what's allowable. There might be someone at the company who's very confident they know what to build but if they aren't aligned with the customer then they're not the decider.

Once we've identified who can say that a feature is or is not built correctly then we measure the distance between that person and the engineers on the team that will build it. Not just the PM on the team, the engineers themselves. The higher the distance the lower the team's knowledge.

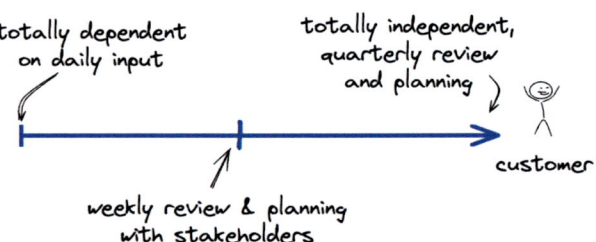

An engineer's knowledge of the customer can be thought of as their proximity to the revenue-generating customer. When an engineer builds something that they plan to use, like a new development tool or a deployment system, they don't do retros with a client; they are the client.

When they're building something for their whole team to use they can still make decisions with their own intuition because they're one of the users. To fill out the rest of the product knowledge they'll need to supplement that with periodic feedback from their teammates.

When they're building software in a domain they don't know at all like, say, tax software for a country they don't live in, they can't trust their intuition about what the product should do. They'll need to tightly iterate with a specialist in that field as well as potential customers. Without both a domain specialist and a user representative they'll definitely fail to build the correct product.

Cadence of Accountability

A team's knowledge of their users' needs determines the level of product support and the cadence of oversight. The less knowledge a team has the quicker their work can go off track.

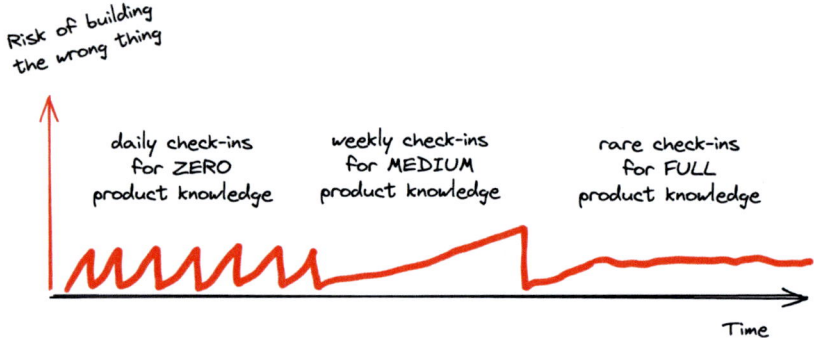

There's a spectrum here between zero knowledge (totally dependent on outside input) and full knowledge (totally independent) that dictates how much the engineers can trust themselves with design decisions, implementation decisions, and schedule. How long they can go without someone with more knowledge of the user checking their work?

If the engineers are brand new to the domain and the customer is completely inaccessible to them then we use weekly (or daily) product-related check-ins with frequent product demos, beta testing, and review meetings. There's no point to a one-week cadence if the team can only see a day (or an hour) ahead before getting feedback from the customer.

Once the engineers become more familiar with the end-user's needs — or if they're working on a product they inherently understand better — then we use a lighter process to give the engineers more time to focus.

And in those rare but wonderful cases when the engineers know precisely what to do then we can confidently create multi-quarter plans with a somewhat infrequent cadence of accountability.

Getting this wrong in either directions adds drag to the project. Too little structure and support means the product is frequently wrong and needs reworking. Too much structure and support introduces drag and the engineers lose their autonomous drive, moving slowly with little creativity.

Pitfalls

There are two major pitfalls worth calling out. One is when a trustworthy senior engineer with low knowledge tries to build something without a design. It's too easy to mistake high skill for high knowledge. Just because an engineer can build anything doesn't mean they'll know what someone needs of them, even if what they're building is an internal or infrastructural tool.

The other is when an engineer claims to have high knowledge of what their colleagues need. This is extremely common for internal tools. No, nobody needs your new metrics library or service framework. This stuff is fun to build but either there are real users (whose names appear in a design doc) whose needs should be centered or this is a side project to do at home.

Levels of Skill

when the team knows this specific implementation — **Expert**

↑

when the team has worked on something similar — **Familiar**

↑

when the team has never touched the technology — **Fresh**

Earlier in my management career I led several teams of veteran infrastructure engineers. These teams needed me to navigate bureaucracy and fly air cover for them. They didn't need much else and certainly not any technical support with the work because we were fixing systems that they themselves had built. This led me to mis-learn some managerial skills and the first time I led a more typical team I had no idea how to make technical support a part of the team process.

I now know that, no matter how impressive an engineer might be generally, they'll need more support when they're less familiar with the technologies involved on any specific project.

The seniority of an engineer changes task-by-task. Someone who's a staff engineer at a SaaS company might be a junior engineer at a video game studio. Someone who works on massive distributed file systems at Google might fall right on their face trying to pre-render HTML using modern Javascript frameworks.

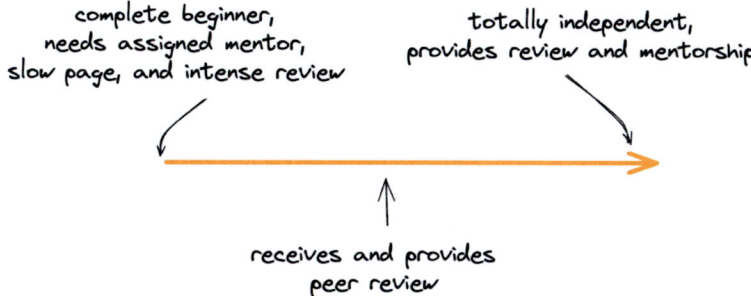

On a scale of low to high familiarity: An engineer could be totally unfamiliar or might have invented the technology in question. There are several qualities about how a manager should structure a project depending on where there engineers fall along this progression.

Technical Oversight

For the purpose of training engineering managers, I break skill levels down into three groups: *Fresh*, *Familiar*, and *Expert*.

Fresh engineers are new to these technologies entirely. Maybe it's backend engineers trying to write a mobile app or a using a language that's completely new to them. These teams need a tight structure that functions like an onboarding tutorial for each technical pieces, ensuring they understand the job before trusting them to succeed at it. No matter what other technologies or engineering problems they might be great at — even if they're expert at some parts of a project like the programming language — if they're unfamiliar with concepts they're about to face they need their manager to plan for and give them specific support.

Even if the team has full knowledge of customer needs their work may still if they lack the relevant skill. The knock-on effects of that stalling is a profound morale hit to the engineers and to all colleagues observing their stalled work.

The strategy with a fresh team is threefold:

First, reduce the tasks sizes so small that the team can feel a sense of forward progress at least every week if not every day. It's emotionally draining to be incompetent at our job and when we face new technology we're (temporarily) incompetent at it. This provides some thrust to get the team moving. Anything to build momentum and stave off the waste of a disengaged engineering team.

The second step is to engage intensive technical mentorship wherever you can find it. If it's available at the same company then the manager should arrange for teaching sessions and group code review guided by an expert. If nobody at the company can help with technical mentorship then the manager should enforce pair-programming. Some of the limited research in our field has found that that's a powerful way for junior people to do highly complex work.[28]

The third is to subject everything they build to design review by other teams at the company. Having one's best write-up of a design reviewed and critiqued by a more experienced person is one of the best ways I've seen engineers grow. It also prevents the low-skill team from introducing technical patterns that will unknowingly compromise future work.

Familiar engineers know technologies similar to this but not this specific implementation. The process here is far more typical. It might take the engineers some time to get moving but with access to people who can answer their questions they'll make progress.

Here, a manager should ensure:

- Design docs are written exhaustively but they're as much for checking one's thinking as getting oversight
- The team has access to engineers with relevant skills at the same company (or another one, if necessary)
- The team demos their work with technical deep dives on a regular basis, at least internally

---

[28] E. Arisholm, H. Gallis, T. Dyba and D. I. K. Sjoberg, "Evaluating Pair Programming with Respect to System Complexity and Programmer Expertise," in IEEE Transactions on Software Engineering, vol. 33, no. 2, pp. 65-86, Feb. 2007, doi: 10.1109/TSE.2007.17.

Expert engineer teams know not only the general technologies involved but this specific implementation. Here, the purpose of our interventions is to grow the team's skill further and to disseminate their skills across the wider org.

A manager should ensure:

- Design docs are to clarify plans and practice thinking through security, analytics, and monitoring nuances

- The team creates post-project write-ups, blog posts, and presentations to disseminate context to company

- The goals are articulated carefully but with wide technical latitude in reaching them

Deliberate Engineering

The End of "Working On"

One of your engineering managers is preparing a review for an engineer they support. They come to you with anxiety about how hard it is to give this engineer feedback. This engineer has been working late and committing many changes to the code. But the manager doesn't really understand what the changes are. Or what's driving the engineer. Day after day this engineer's updates are that they're "working on" something but it's not clear what progress is happening, if any.

It's possible this engineer is saving the company from catastrophe every day. It's also possible they're reformatting files and making unnecessary tweaks in low-value areas. The manager doesn't know, and maybe adjacent engineers are too busy with their own work for them to be much help to the manager in figuring this out.

It's possible this engineer needs the freedom to make changes only they can understand. It's also possible they're absolutely spinning their wheels and need to be redirected toward real work. It's hard enough to figure out the answer for one engineer but it's much harder to figure out the answer for all of the engineers in your organization.

Promoting based on 'impact'

Some companies have found a shortcut to this problem. They promote and reward engineers based on the visibility of their work. They might call it 'impact' but if the impact benefits from a flashy demo or simple metrics then it's actually visibility.

If you're working on something invisible and important then you have my respect but you won't get promoted at a place like that. Find a way to communicate the value of it or else the company will assume there isn't much. This leads very quickly to gameable promotion systems where getting a big raise depends primarily on project selection. All you need to do to get ahead is "drive" some "impactful" projects that "deliver" business value. It's a simple script you can follow if you have the right political insight.

The end result of this is an engineering culture with incentives tuned toward flashy launches and a rapid accumulation of high-interest debt. Maintenance is undervalued, as are incremental improvements. You may be able to think of a specific real-life large company with a habit of launching products that it shuts down soon after, often launching multiple big, redundant products in the same space. This happens when we incentivize flashy work.

There's another shortcut to this problem that doesn't cause so much damage. In fact, it improves the careers of engineers and streamlines both their growth and the effectiveness of their contributions to the company.

The shortcut is eliminating the concept of "working on" something. We don't want to pay a car mechanic to "work on" a car, we pay them to fix a problem with the car. We don't want to pay a teacher to "work on" being our teacher, we pay them to improve our knowledge and skills.

The Three Meaningful Activities

To avoid the proliferation of "working on" activities, I've found it clarifying to set a policy that there are only three useful types of technical work:

1. Design (or redesign) a system

2. Bring a system in line with its design

3. Incident Response

These are mutually exclusive and comprehensively exhaustive with the technical work you need your engineers to do. Any other technical task is of dubious value or needs to be allocated within the L&D budget because it's for personal enrichment or it's a waste of absolutely everyone's time.

There's plenty of other non-technical work you should expect in the form of mentoring, training, improving organizational equity, personal study, recruiting, interviewing, representing the company, etc. But of the direct technical work all of it can and should fall into one of these three categories.

The reason this list of three activities is so powerful is that it captures and elevates the *intention* of the work. You're paying people to build software for a purpose. Only (and all) of the software that fulfills that purpose should be created. It should be designed for a purpose and it should be implemented as designed, even when the designer and implementer are the same person.

Activity 1: Design (or redesign) a system

A project manager is friends with a junior engineer. The PM comes the engineer with a request, asking if it's possible for any prospective user who visits a landing page to immediately appear in the admin UI under 'users' rather than wait until the user is fully-onboarded. The engineer says "yes" because they want to be helpful and they know they can do it. This is computers, anything is possible.

What impact does this decision have on existing features? On future features? What problem does it solve? Which problems does it create?

Technical design is a scary process for some junior engineers because they imagine that one needs to become extremely senior – maybe even an "architect", whatever that means – before they can do effective design. It's true that as we gain experience in more systems we have more patterns to reference, but this is merely an optimization. Very junior engineers design solutions to problems as part of their daily work, even if they don't realize it.

The technical design process is not about making something beautiful or interesting. It's certainly not about making it complex. And it's absolutely not about inventing something clever.

Technical design is merely documenting our thinking as we to solve a problem so we can see our mistakes early. It's how we take maximum advantage of that space between recognizing a problem and solving it – a moment where we can consider our options and collaborate with peers. Ideally, it gets us toward a solution that's maintainable, secure, observable, operable, and low-cost. But even if it doesn't change what we would have built it serves as a process of thoughtful organization and future documentation.

For engineering cultures that don't do written design the process may seem to provide dubious value. But as soon as there's even one document describing why something was built the way it was the experience of future engineers is improved. They'll be able to see not just the implementation of something but the original point of it, and be able to evaluate if it's still the best solution for the problem it addressed.

The best technical designs I've seen are in design docs (or "tech specs") and contain a simple (if sometimes verbose) strategy. They start with a problem statement, with measurements if possible, followed by an exploration of possible general technical approaches to solving it. There's some explanation for why one approach is best and then a structured description of how this approach will be implemented.

The most important thing in the design stage is giving the engineer(s) writing it a chance to reflect on their thinking. By articulating our thoughts we find their errors.

The second most important thing is getting other engineers to read it in detail and understand it. Not just for their feedback to improve the plan, but primarily for their education. I know no faster way to advance in one's engineering skill than reading designs very carefully and trying to improve them.

There are many templates for this kind of document[29] but any template that prompts the author to think broadly will work. The two most important questions are "What problem is this solving, exactly?" and "What alternative solutions have been considered?"

That's the first of three meaningful technical activities. The design phase might feel like mere paperwork but the value is clear once we get to the second meaningful activity.

Activity 2: Bringing a system in line with its design

This same project manager asks their engineer friend about a feature. They want the user signup flow to send emails directly to this project manager every time someone signs up.

Can the engineer do this? Sure. Should they? Not unless this is the Business Intelligence plan for user analytics. Which it absolutely should not be.

---

[29] I use https://jackdanger.com/design-doc-template

If the user signup flow has gone through any kind of design process then there's a better, maintainable way for the product manager to get this information or there's a big blank area under an "Analytics" or "How will we observe user behavior?" section of the design and anyone who signed off on the design doc should talk through how that happened.

The engineer here should be able to say "No, I can't do that because that's not how we're handling user analytics. But I can give you access to this dashboard we're making as part of the onboarding system." A little up-front design helped the engineer avoid adding ridiculous amounts of technical and analytics debt while also getting the project manager exactly what they need.

When a design is absent, go through the engineering design process. When a system is broken (or there's a request for changes) then examine whether the design has been incompletely or incorrectly implemented (in which case finish the implementation) or whether the design itself is incorrect or incomplete. If so, return to the design phase and then bring the system in line with that design.

Activity 3: Incident Response

Virtually all of the technical work engineers do is captured by designing and implementing. The one exception to this is when the unexpected happens.

This same product manager comes to an engineer and announces a major bug: The signup analytics dashboard doesn't have any data on it. Maybe this dashboard was meant to have no data and, weird, but sure. We can question the design process that led to that choice but it's not a bug.

Or perhaps the feature is simply not done yet, in which case the engineer can give a rough estimate for when it'll be done.

But maybe the dashboard is supposed to have data and this is a regression. If that's the case then it's time for *incident response*.[30]

---

[30] If you're not familiar with incident response search online for "crew resource management" and check out the decades of study that firefighting, surgery, and aviation have put into defining this discipline.

Incident response is reality's megaphone. It's where the facts of the computers get so loud they drown out our mere opinions and plans.

Incident response is how we deal with failures. Not incomplete features, not poor designs, and not executive whims. Those are all unfortunate but expected. Incident response is a process that's used explicitly just for when we find out something true about our system is, in fact, false.

This isn't just a process for big companies, either. Even at a startup with only 3 or 4 engineers this process is essential because it allows for clear thinking about whether the fault lie in the design, the implementation, or if there's any fault at all. An early-stage startup can't afford to make everything work, much less all work perfectly. So it's critical to use an incident response process as a way to figure out if some error or complaint is a regression of a necessary feature or just something that would be awfully nice if it worked.

Turning This Into a Policy

The easiest way to use this pattern is for an engineering manager to simply ask an engineer which of these three they're doing. In a high-trust, calm environment it's not rare for the engineer to admit that they're just spinning their wheels and feeling lost. The manager can then work with the engineer to diagnose which of these three are the important next step, unlocking the engineer with a clear goal: Either write/edit the design or finish implementing it. Often both, in that order.

Done that way, this is a cultural change. Which is deeper and longer-lasting than a mere policy but also much slower to take hold.

To get the benefits of this quickly it needs to also be a policy; a scheduled and written practice that the organization does. The lowest-friction and simplest implementations of this that I've seen are:

- Clarify in the template for the employee review cycle that the only technical work that counts is these three

---

Incident Response, in brief, is the process a team goes through to identify a crisis, stabilize it, and then buy time to calmly fix it later all while minimizing damage, accelerating learning, reducing systemic risk, and improving morale.

- Instruct engineering managers to modify standup meetings for 3-6 weeks where every IC must say which of these three they did the previous day and which they're planning to do today

- At the next engineering management meeting identify which ICs are most prone to non-output oriented busywork and make a plan to dig into that work. The result should either be that the company discovers the IC is working on a critical, hidden problem that leadership needs to plan around or that the IC receives more directive, supportive coaching toward more valuable work.

Communication Design

"There's no information and all of it is wrong"

— half the people we work with

At every company I know there are complaints about information tools. Email is an abandoned wasteland, the wiki is incomplete and incorrect, there are too many channels but none of them are right, etc.

Poor communication architecture is a massive source of drag, exposing people to the friction of noise while denying them information they could move them forward.

It doesn't have to be this way and you can fix it.

The early drafts of this book didn't include this fourth team empowerment. I thought it was too confusing to try to shoehorn communication system design into a book about lowering engineering drag.

But as I reviewed my notes I saw that every struggling org I've worked with was in a context where highly paid engineers didn't have the right information and, crucially, their leaders didn't either. A prerequisite to addressing even basic problems is helping everyone communicate about it.

You can immediately improve the effectiveness of your org through internal communication design.

There's that old saying "If you want to go fast, go alone. If you want to go far, go together." Deciding how to make that tradeoff is hard, and depends on situational context.

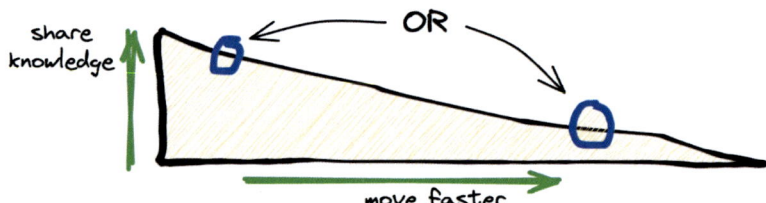

On one end we do our work quietly. We avoid distracting each other and we move at full speed — in totally different directions.

On the other end we share everything and sync constantly. We've achieved maximum alignment but the sheer noise slows us to a halt.

We design internal communication to avoid this tradeoff entirely.

Our goal is for everyone to have the information they can use while not slowing down either the people producing that information nor the ones consuming it.

Before doing that we have to address our communication values. Specifically, what level of transparency is appropriate for our organizations.

When I joined Square it was the first transparent organization I'd experienced and it shocked me. The shock came in two forms: How will I make sense of all this information? And why is there suddenly so much scrutiny on *my* work?

The shock faded quickly and the benefits were extremely clear. In the same way that an eight-person team can do things that eight individuals can't, a large connected org can do things that all the individual teams can't.

But we need to get people past that shock I first felt: How do I consume this information and what's a good way to produce it?

The next section has good answers to both, but as leaders we must deal with another critical detail: Does everyone in the company get visibility into *our* work?

The quickest way leaders can ruin a good culture is to not participate in it. People follow our lead, by definition.

Therefore, we need to decide up-front how much transparency into our work we'll offer. Personally, I approach it this way: If a decision might affect a colleague's work or a customer's experience then it must be broadcast. And if a discussion might lead to a decision like the above then it must be broadcast as well.

Not just available on request, but broadcast.

I remember the Square executive team sending out weekly notes of their meetings to thousands of employees leading up to the (confidential) IPO. Anything sensitive was redacted but no employee lacked access to high-level context on the company's focus at all times. Hundreds of other teams did the same.

If a financial company can be that transparent during a time of IPO preparation then anyone can.

(You might think this kind of massive information broadcast wouldn't scale but it worked from a hundred employees well into the thousands. Square was founded in 2009 with one prototype of one product but went public a mere six years later with twenty-four different products. There are many reasons for this incredible product success but a major contributor was that literally all of the work at the company received scrutiny and collegial support from everyone.)

Better Decisions, and Access to Feedback

To be a little more concrete, the two things we're aiming for with our upgraded transparency are better **decisions** and access to **feedback**.

A company I worked with was talking to a vendor to buy a certain hosted massively-scalable message bus. The kind that could process trillions of messages every minute.

All stakeholders were on a call when the vendor's solutions architect asked which tier of service we needed.

"How many messages per second do you currently send?"

The answer, from our side, was "two hundred and forty."

A long pause.

Then, from our side again, "sorry, that's per minute. So... 4 messages per second"

A longer pause. Some of the ambient noise disappeared from the call – someone had muted to laugh.

Afterward, we met up on our side to talk about how, exactly, we had gotten into this situation. The further we looked the less we could find any paper trail of decisions around this vendor, around the message bus in question, or around any kind of reasonable problem statement. We'd somehow picked the largest possible scale tool for a problem that could be solved with literally any technology. A modern doorbell can handle 4 messages per second.

This is one of the two major pitfalls of an under- or poorly-designed communication system. If decisions aren't written and broadcast at the time they're made then stakeholders can't know why they were made, they can't improve them, and later on nobody can find the context to explain why we're in a predicament.

The other major problem from poorly designed communication systems is a lack of feedback.

Feedback requires a feed. For all the training sessions on feedback that focus on how to give feedback, the original feed is really underemphasized. There's no way to respond to something we cannot perceive.

If an executive makes announcements once per quarter at an all-hands meeting and says "If you have any questions don't hesitate to reach out" they're unlikely to get a response. If that same executive had been publishing their decisions and strategies along the way they won't need to prompt for feedback when they're holding the mic – they'll have received it along the way and the company plan will already be improved because of it.

If an engineer asks their manager for feedback after having given vague status updates for several weeks they're equally unlikely to get much. If the manager is experienced they may force the engineer to produce a feed with a question like "What would you like feedback on?" but even then we can't sculpt something big when we're given small amounts of clay.

So we want to make better decisions and we want a culture of better feedback. The way there is to design a communication system that gives us both of those while empowering our people in their work.

Designing a Communication System

Our colleagues may complain that there's too much communication but the problem is likely there's just too much noise relative to the signal.

As an executive you can drive this signal to noise ratio very high, empowering your teams. You do this with policies for how your org communicates. Even if you only lead a small part of the company (or you're an IC and you can only influence your own team), if you create an effective comms system then the rest of the company can adopt the parts of it that look useful.

We maximize the signal to noise ratio of our communications by putting useful information where people can make best use of it, without dragging them through information they can't use or by forcing them to expend much energy producing that information.

With the above in mind, let's get into the details. We have to choose technologies and policies for those technologies. Our goal isn't to use one technology better, it's to rebalance our use of all technologies so that each one feels easier to work with.

There are nine media we have, each with tools perfect for just one medium:

1. email
2. group chat/instant messaging
3. documents
4. slide decks
5. spreadsheets
6. wikis/knowledge bases
7. task trackers
8. video meetings
9. source code

That's it. That's nine places to put information, each of which is useful for a different purpose.

We keep drag low by using each tool for only the things it is uniquely best at. No tool that's excellent in one of the above categories will do a great job in one of the other categories.

Most of these tools have a plugin architecture that's Turing complete so, yes, they can each do everything. And there are some well-funded campaigns by a few big companies (e.g. GitHub, Slack, Salesforce) to convince us to use their product for all information.

Despite this, each product only really excels at one use. So instead of putting all the information in one place we'll put each in the one right place.

Source code systems are low-drag for anything a computer executes.

Spreadsheets are low-drag for any tabular data.

Slide decks are low-drag for presentations.

Video meetings apps are low-drag for talking to each other outside the office.

What about group chat, then? At the time of this writing many startups have opted to shove all of their information into Slack. It frees users from having to use multiple tools. And then they immediately leave/mute/ignore the proliferation of channels and messages because a tool for instant messaging turns out to be good mostly just for instant messaging.

Turning this into a policy

Designing a communication architecture has some nuance because information is so complex, but we can get really far by trying to do just two things: Make sure the information is produced and make sure it's put in the right place.

We'll start with creating a policy for producing decisions, then look at the individual tools.

Recording decisions

The most important policy here is around publishing and recording decisions in a way that is accessible to the whole company.

Personally, I prefer an email-based workflow for recording decisions. The policy I like is the following:

```
All work meetings (not social events or interpersonal
1:1s) must publish meeting notes to a company-subscribed,
archived email list.

Any decisions made in that meeting must be communicated
under a header called 'decisions' so someone can search
for that word in combination with other terms. Any
decision not recorded here is not decided.
```

I've seen similar policies done with a Slack channel called #decisions, where emoji reactions help with discoverability by annotating the impact and scope of the decisions themselves. I've also seen this attempted in wikis but never for long due to the high drag of publishing.

Pick a workflow that feels right and iterate as you go - the important thing is that the decisions get made, that they get the benefit of feedback from anyone who might be able to provide perspective, and new hires can find historical ones.

Beyond that we'll use each tool for just what it's best at.

Reviving your Inbox

Before talking about tools we need to face a hard truth: Many people have abandoned their email inboxes. When I work with junior engineers they often don't use email much (or at all) in their personal lives and have let their personal inbox become a wasteland ransacked by marketers. On their first day at the job they see that bots and other automatic notifications are in their work inbox as well and assume that this, too, is a lost cause. That email itself is a legacy technology as useful as jQuery, Fortran, and vacuum tubes.

The first step for us as leaders is establishing the expectation that each of us will be responsible for knowing the information that colleagues email to us. And therefore that we're responsible for the hygiene of our inbox.

The second step is helping people actually do that. Fifteen minutes of filters can turn an overgrown email wilderness into a curated garden. But knowing how to do that requires either years of practice or, more helpfully, a one-hour class for all new hires showing them how to do it with the specific email technologies your company uses.

An old colleague[31] used to do precisely this to every group of new hires at Square. Within an hour he had them set up with filters and an approach that would amplify their work for the duration of their employment.

I recommend starting simply by having all your colleagues create a filter that bypasses the inbox for messages with the string 'unsubscribe' in them.

The next step on our journey is creating the right expectations around when we use email and when we use instant messaging. Both are very powerful. Both are easy to abuse and terrible places to try to stuff all content into.

---

[31] Xavier Shay, who I cannot recommend strongly enough for literally anything.

Email for broadcasting non-urgent information

Email is a great medium for broadcasting information that isn't urgent. I define non-urgent as "can be read any time in the next two working days."

Email messages, no matter the size, have a subject line that even the pathologically vague will try to put something into. They also have enough room (effectively unbounded) that it's possible to use the Inverted Pyramid[32] where all the content is summarized in the first line, then there's a paragraph summarizing it in more detail, and then the rest of the content.

Email allows adding new individuals or groups to conversations as well as forking a conversation by removing some recipients. Two features that empower communication and are – as of the time of writing – wholly absent on all popular instant messaging platforms.

Done right, it's a way for all colleagues to quickly skim and archive huge amounts of information. Done poorly, it's still a better place for long-form text than instant messaging.

Email is also a great medium for long-form discussion. Something that isn't urgent at all but is very intricate. This is how we get technological marvels like Linux.

The deepest engineering work I've ever done had nothing to do with writing code. It was long mornings at the office just sipping coffee and very slowly reading a back-and-forth thread of write-ups by multiple colleagues on specific problems we were about to encounter the following year with our data scalability.

Had this happened in an instant message context then other messages would have blasted it off the screen or it would simply not have fit.

The ergonomics of instant messaging do not welcome deep thinking.

---

[32] This pattern is generally recommended. The format of an online recipe – putting the actual recipe below a long story – is to be avoided in work communication.

Instant Message as a default

What instant messaging does right is instant communication. And as a default it lets people ask where another, better place is to publish or find information. Which is hugely powerful when the structure makes sense. But out of the box no company's instant message structure makes sense.

The painful parts of instant messaging at work are that there's either no channel for what we're trying to say, or the appropriate channel serves multiple purposes.

Our goal here is to create one obviously correct place to have any kind of conversation, with mostly just that conversation in that place. This is a departure from a common anti-pattern where channels are grouped around teams or projects. Noise is drag, and having to converse in a channel where unrelated and uninteresting conversations are happening is very noisy.

So the policy here is: Create channels for specific types of conversations, welcoming all who want to participate or observe

Measuring the ratio of public to private conversation can help us find where people are struggling to communicate in the open. When colleagues use direct messages for work those conversations are totally opaque. There's no chance for colleagues to learn, decisions to be improved, psychological safety to be increased, or managers to coach.

When I ask colleagues what would help them avoid direct messages they give one of two answers: Either they were embarrassed to ask a question in public or they didn't know where to have the conversation.

If it's the first, this is our opportunity to increase psychological safety. A company where you can ask a silly question in front of everyone and trust to be treated with complete respect is a high-performing company. We, as leaders, need to push these questions (and ask them ourselves) into public spaces so we can find and eliminate any sources of shaming or blaming.

If it's the second — people just don't know where to discuss something — we need to make a place.

And it should be a public channel. There's a school of thought that private channels should be prohibited at work. That might be extreme, but private channels should probably only exist when there's an articulable reason for the privacy.

One reason for this is that most messaging apps don't even tell people which private channels exist unless they're in those channels. So there's a high chance somebody could be passively learning or maybe even actively helping a team's work but won't, because they don't know that work is even happening.

Standardizing Channel Types

But if everybody is talking in the open and there's a channel for everything then that's overwhelming! How can we get work done if just anybody can drop in and talk to us everywhere?

The best solution I've seen for this, unsurprisingly, applying yet more design. We can create a room for each kind of conversation a team will have. If there's a specific channel for people to talk to your team then they won't talk in the one where your team is working.

One specific pattern that I've seen work very well gives each team at the company four specific channels with predictable names:

- `{teamname}-help`
- `{teamname}-core`
- `{teamname}-lounge`
- `{teamname}-alerts`

The `-help` channel is where people from outside the team can come ask any questions at all.

The `-core` channel is a workshop for the team and any stakeholders who are shoulder-to-shoulder with the team

The `-lounge` channel is for all the fun stuff that a team needs in order to bond, in a place where they don't have to worry about jokes interfering with work (I'm okay with this being private but as a public channel it can be a source of joy for the whole company).

And `-alerts` is a way to keep robots off to the side to avoid ruining the vibe.

This is the bare minimum channel architecture but it's enough so that anyone can talk to any team about that team's work and trust that they are welcome to do so in public. They just type in the team's name, find the 'help' channel, and ask their question with confidence.

Wikis

Wikis are super powerful and as of this writing there are several new, extremely sophisticated content systems that work well for this.

A wiki, like any other piece of technology, can technically hold any kind of content. But it's a terrible place for anything time sensitive. Many of the modern systems would happily embed a project tracker for you in the wiki but there are other tools better suited for this (i.e. project trackers).

A good wiki policy is simple and has two parts to it:

1. All employees must be able to edit any page of the wiki

2. Every team must maintain at least one page in the wiki to explain their team to the company

These are super easy to do but surprisingly rare. I've now been in two long conversations (one spanning months) attempting to talk an HR team member into allowing edit rights to the wiki. In both cases the moment the wiki was unlocked the value of it skyrocketed.

And a good team onboarding wiki wasn't important to me until I saw a great one for the first time. Not only did it have the team charter but it contained tutorials to help both new hires onboard to the team and help other teams understand and make sense of the team.

It's a powerful tool not only for smoothing out the team's work and for onboarding new teammates, but it sets a very high bar for professionalism that then bleeds into all of the other work.

Documents

Let's not forget documents, the workhorse of collaborative communication. An entire other book could be written on document-oriented workflows. For our purposes here I'll just mention the high-level strategies that reduce drag on Engineering.

The meeting is in the document

The meeting may have a conference room or a video call but that's not where the meeting exists. The meeting exists in a document.

The field of engineering has a high rate of neurodiversity and this policy helps us support our neurodiverse and disabled colleagues while also improving the effectiveness of our teams.

Maybe someone is socially anxious and can't speak up in a meeting or can't read and think while someone else is speaking. Or they're Deaf or otherwise use assisting devices to speak. In all these cases you can count on them having a way to participate in a document.

The policy "The meeting is in the document" centers the document in a way that allows for chats, video calls, and other interactions to orbit the document in whatever way works for people. As long as the content of those interactions somehow makes it into the document then the interactions are an effective part of the meeting. In this way we can cater to people who have to process out loud as well as everyone else.

The positive side-effects of this policy are vast:

- Everything is documented for future review

- Participation can happen concurrently, massively increasing the bandwidth of the meeting

- There's a full edit history for us to check our own work

When I'm going on vacation but my teams center their meetings in documents I trust that I won't miss anything.

And if someone in my org holds a meeting but there's no document (or, just as bad, no notes were sent out) then I won't hesitate to schedule a repeat of the meeting and make everybody do it again, better. We set a high bar for our work and we don't leave anybody behind.

Don't Bogart the cursor

A simple but powerful change: Make sure there are extra bullet points below the person who's typing.

It's fine to have a single note-taker at a time but when the meeting is happening faster there is more happening at once. Rather than block concurrency, let's embrace it. We do that by designing a place for that concurrency to happen.

So, when someone is typing and they're typing on the last line of some bulleted list, make room for someone else to type below (or above). We don't Bogart the cursor.

Meetings begin with silent readings

And the final thing I'll mention on documents is that a silent reading at the beginning of a meeting is vastly underrated. Many leaders (myself included) have a tendency to try to jump into discussion but that discussion will be better if everyone already has all the information in their heads and has had a chance to consider what they think of it.

If a legible, informational document can be prepared ahead of a meeting then a great use of time is to spend the first part of that meeting — even a majority of it — silently reading. Even if the discussion afterward is brief it's more likely to yield thoughtful, correct decisions. Jeff Bezos famously did this with his senior leadership team to great effect.

You may choose to allow document comments and silent discussions during some or all of this portion of the meeting. If you do, you might want some way for people to signal that they're done reading. The cutest approach to this I've seen is to create a "cursor garden" somewhere in the doc with lots of flower emojis. Once someone parks their cursor in the garden you know they're done.

Empowered Thinking

The point of all of this is to maximize the creative power of your highly paid team. As they say in Data, "garbage in, garbage out." If your engineers have less than a full picture of the whole system you'll see it in the way they collaborate.

There's no right way to design a communication system because, like most design, it's unique to a problem and the materials on hand. Use each tool for its strengths and organize the information flows such that people can predict where the information will be. Doing so unlocks the power of your senior people by giving them all the information they can use while keeping your junior folks effective by not burdening them with noise.

Applying all four empowerments

As executives we're so far from the work that our points of leverage are quite abstract. If management is the derivative of work then managing managers is the second derivative. Where does that leave an executive, with anywhere from two to five layers of indirection between them and the work?

An efficient approach to concrete improvements is through the four areas I just outlined in this part of the book — giving our engineering teams **impact**, **process**, **focus**, and **information**. These empowerments set up each team to do its very best within the larger context we defined with Technical Coherence.

Airborne

Your company might run out of money.

This is a foundational truth of business and a reason for your CEO's restless nights.

Luckily, your CEO has you. And you know that engineers are a source of miracles. The ingenuity and insight of your engineers can deliver your company to the loftiest heights as long as they're given the right context and leadership.

You can identify realistic measurements of velocity and drive them higher over time.

You can apply the 3 steps of Technical Coherence to streamline your organization into a shape that takes you where you need to go.

You can plan the route ahead in detail, navigating your company deftly around unnecessary debts and paying down some before they could slow your pace.

And you can ensure each engineering team delivers maximally to the cause, no work wasted, no toil dragging on morale, and no engineers wondering who they're supporting or how its going.

The job of an executive is hard. This is true in any function, and Engineering has abstractions and complexities that make it even harder. We make it easier for ourselves when we find alignment between us and our CEO, between us an our CPO, and between engineers and the rest of the company. With that alignment everyone in Engineering can do the right thing for the company and also the right thing for their team.

And your whole organization can take flight.

A favor

I've put my head and my heart fully into this book, hoping to make it useful to you. If you leave a review online right now I'd appreciate it deeply. I wrote this on my own and self-published, so I'm relying on you, my reader, to help get the word out to everybody who could benefit from this.

With all my thanks,

Jack Danger Canty

https://jackdanger.com

About the author

Jack Danger has a silly name, a loud laugh, and a lot of pride about having finally written this book.

He was raised as a country boy in the foothills of the Cascade Mountains by his mother and grandmother. He taught himself to code on a PC he bought with a below-minimum-wage pizza job. Despite loving programming, he couldn't see how it would fit into a satisfying life.

He was the first in his family to go to college and he paid his way by working on startups (and Pell Grants — hell yeah Pell Grants).

In his mid-twenties he met his father and father's side of the family. His father can make anything out of metal and his grandfather could do the same with electrical circuits. His grandfather worked on the world's first realtime digital computer at MIT and spent a career at MITRE and Boeing leading the AWACS program which produced the Being E-3 Sentry pictured on the cover of this book.

Jack is grateful to have found his place in this engineering tradition.

Acknowledgements

Many people reviewed early drafts of this work and improved it substantially. My biggest thanks go to David Haslem for his multiple thorough reviews and to Lydia Thornton, Filipe Giusti, Miriam DeAna and the entire Pathstream engineering team for incredibly helpful early guidance.

Thanks to Kevin Davis for helping me remove useless parts and for making the airplane metaphor sillier, to Will Larson for modeling the engineering leadership I aspire to, and to Uma Chingunde for talking me out of my worst ideas.

To Smruti Patel for providing detailed full-book feedback alongside some of the best career coaching I've received.

To Davy Stevenson, for working alongside me as we learned how to be executives in companies with too little time and money for us to grow the slow way.

To Carlos d'Avis for his crystal-clear focus on the important parts of engineering leadership.

To Wenley Tong for his razor-sharp insight, as always.

To Ryan S. for the kind of deep feedback that forces a partial rewrite.

To Max Countryman for helping me talk through the whole book.

To Jennifer Chao for demonstrating the kind of leadership this industry needs.

To Megan Dawson for her eagle-eyed editing and helpful early feedback.

And most of all to Ruth Dawson for helping me finish this book about my silly industry as you work a real job protecting our democracy.

Printed and bound by CPI Group (UK) Ltd, Croydon, CR0 4YY

18/03/2024

03742627-0001